PUBLICATIONS OF THE ISRAEL ACADEMY

OF SCIENCES AND HUMANITIES

SECTION OF SCIENCES

———

On the Structure of the Spermathecae and Aedeagus in the Asilidae and their Importance in the Systematics of the Family

On the Structure of the Spermathecae and Aedeagus in the Asilidae and their Importance in the Systematics of the Family

by

OSKAR THEODOR

Jerusalem 1976

The Israel Academy of Sciences and Humanities

Printed at Litho-Offset Ziv, Jerusalem, Israel

CONTENTS

INTRODUCTION

THE STRUCTURE of the spermathecae has been found to give important systematic characters in some groups of insects. The spermathecae have been used for a long time in the systematics of fleas and have provided one of the principal characters on which a revision of the Phlebotominae was based (Adler and Theodor, 1926; Theodor, 1965).

During a study of the regional fauna of Asilidae and Bombyliidae, the spermathecae were examined and a surprising variety of forms was discovered. Examination of the aedeagus and other parts of the male genitalia of Asilidae showed that there are characters which have not been used at all for systematic purposes; certain parts have been described rarely and not in adequate detail, e.g. the form and chaetotaxy of the gonopods and the proctiger of the male. Some of these characters are apparently of generic rank, but specific differences were found in practically every case in which several species of a genus were examined.

The literature contains little information on the spermathecae of the Asilidae. The spermathecae of *Laphria flava* have been described by Reichardt (1929). Owsley (1946) described the spermathecae of several American species, and he mentions descriptions of the spermathecae of some species by Loew, Dufour and Sturtevant. The spermathecae of the Bombyliidae, Nemestrinidae and Mydaidae have apparently never been described. Only a few incomplete descriptions of the spermathecae of Tabanidae have been published (Mackerras, 1955; Oldroyd, 1954; Ovazza and Taufflieb, 1954).

The male genitalia of the Asilidae have been widely used in systematics, but most of the descriptions give only their external appearance in pinned specimens, and if the aedeagus is mentioned at all, only its outline, as a rule of the distal end, is described.

The only detailed description of the aedeagus of three species has been given by Reichardt (1929). Hobby (1936) illustrated the outline of the aedeagus of a number of Ethiopian species of *Promachus* and suggested using the aedeagus in the descriptions of other groups of Asilidae. Karl (1959) gives drawings of the aedeagus of a number of species, and Martin (1968) of over a hundred species, but again only the outline is given and the drawings are reduced to such an extent that details are not recognizable. Weinberg (1961; 1967) used the endings of the aedeagus as a specific character in some species of *Machimus*, and Tsacas (1968) gives drawings of the aedeagus in his revision of the genus *Neomochtherus*, but they too lack sufficient detail.

In the past the classification of the Asilidae and the Bombyliidae was based partly on morphological characters (form of the head, antennae, sternite 8 of the male, ovipositor of the female, wing venation, etc.), but also to a large extent on external characters like coloration and chaetotaxy. These characters have proved unsatisfactory in many cases, particularly in the grouping of species, e.g. in the genus *Exoprosopa* in the Bombyliidae and in the genus *Apoclea*, as well as in the *Machimus* group of genera in the Asilidae. The colour of parts of the body, particularly of the legs and of the setae, has been found to be very variable, so that the same species may show a different coloration in different localities

1

and habitats (see *Saropogon*, species no. 7, p. 90. This applies especially to the extent of black pigmentation in arid and humid habitats.

Wing venation has also proved to be highly variable in many instances, e.g. in the genus *Apoclea*, in which cell r_5 may be open, closed or stalked in different specimens of the same species, sometimes even in the two wings of the same individual. Engel's key to the species of *Apoclea* (1930), which is based partly on this character, thus becomes virtually useless for identification. Another instance is the venation of *Glyphotriclis*.

The form of sternite 8 of the male has been used by Hull (1962) as a generic character to distinguish the genera *Epitriptus* and *Tolmerus* from *Machimus*. Examination of the genitalia showed that *Machimus atricapillus*, which has a long, bifid sternite 8, closely resembles *Tolmerus pyragra* in the characteristic form of the aedeagus and dististylus. Similar differences in the form of sternite 8 were found in *Efferia*, so that this cannot be used as a generic character.

External characters like coloration and wing venation give valuable characters in general, but they are not sufficient to distinguish species in certain genera or groups of genera of Asilidae, in which they either are highly variable or differ so little that definition of species by external characters alone is not possible. However, the genitalia give distinct characters, e.g. in the genera *Apoclea* (Figs. 398–414), *Habropogon* (Figs. 79–93) and *Stenopogon* (Figs. 58–62).

The present paper attempts to show that the detailed structure of the genitalia of both sexes contains characters which have not been used in the past or have not been used in sufficient detail, and which make it possible to distinguish species and in some cases to define their systematic position with more accuracy and with greater certainty than external characters alone.

Eighty-five genera, mainly Palaearctic, and about 260 species have been examined, mostly from dry material, but fresh material was used as far as available for the study of the function of spermathecae and aedeagus. The information from fresh material made it possible to interpret details in preparations from dry material with greater precision.

The undescribed species mentioned in this paper will be described in a volume on Asilidae which is in preparation for the *Fauna Palaestina*.

The genitalia of the Bombyliidae and other families of Brachycera will be described in another paper.

I am grateful to Dr J. Decelle, Tervueren; Prof. J. Kugler, Tel Aviv; Dr F. Kuehlhorn, Munich; Mr H. Oldroyd, London; Dr L. Pechuman, New York and Dr R. Wenzel, Chicago, for providing me with material for my studies.

I would like to thank the Israel Academy of Sciences and Humanities for supporting the publication of this monograph, and the Publications Department of the Academy, especially N. Schneider, for their help in preparing the manuscript for the printer and seeing it through the press.

TECHNIQUE

THE SCLEROTIZED parts of the male genitalia of Asilidae rarely extend anteriorly beyond the last two segments of the abdomen, but in some genera, e.g. *Alcimus*, they extend further anteriorly, so that the posterior half of the abdomen has to be examined. The genitalia are macerated in cold 10% KOH for about 24 hours at 30° C. The parts are separated and mounted in Canada balsam.

Aedeagus: The aedeagus is mounted in lateral view, and also in dorsal view in cases in which special differentiations are present, if only its distal end. If the musculature has to be studied, the genitalia are fixed in Bouin, and removal of the outer parts usually permits us to obtain the necessary information. If the gland of the aedeagus and its inner musculature has to be studied, the aedeagus is fixed in Carnoy and is treated like a section. Dry specimens are kept in distilled water for 24 hours and are then stained with eosin.

Spermathecae: Dry Material. The spermathecae of many species are situated in the 2–3 posterior segments of the abdomen. Those of some species are extremely long and extend far anteriorly, in some instances throughout the whole abdomen, and it is advisable to treat the whole abdomen if the general arrangement and position of the spermathecae in the genus concerned are not known. The abdomen is macerated in cold 10% KOH for 24 hours at 30° C. The contents of the abdomen (eggs, etc.) are removed by applying gentle pressure, but care has to be taken not to dislocate the spermathecae. If the spermathecae are delicate, not strongly sclerotized, they are stained, after washing in water, in diluted haematoxylin or Mayer's haemalum for 20–40 minutes. This makes the canaliculi of the gland cells clearly visible; they are sometimes difficult to see in material treated with KOH. The abdomen is opened along one pleura to facilitate penetration of the stain. It is then passed through alcohols and dissected either in the clearing agent or in the mounting medium. The abdomen is opened along the pleural membranes, the remaining body contents (tracheae, etc.) are separated from the integument, and the spermathecae disentangled. Care has to be taken not to tear the often delicate and long ducts from their base. When the connection with the furca is cleared, this is removed from the ventral part of the ovipositor, and the furca and the spermathecae are mounted together with the last segments, which are of characteristic form in many species.

Spermathecae: Fresh Material. The abdomen is dissected in saline and the spermathecae are transferred to a drop of clean saline. They are fixed under the coverglass with Carnoy's fluid by drawing the fixative through with a strip of filter paper. More fixative is added and the coverglass is floated off. The spermathecae usually remain attached to the coverglass and are kept in the fixative for another 10–15 minutes. If they remain on the slide, fixative is added to it. They are stained with haematoxylin, passed through alcohols and mounted. This technique permits study of the gland and musculature of the spermathecae, and many other details are more distinct than in material treated with KOH.

Spermathecae

Reichardt (1929) described and illustrated the spermathecae of *Laphria flava*, including the histology of the gland. He also described the spermathecae of *Rhadiurgus variabilis* (as *Asilus trifarius*) and of *Machimus atricapillus*, but they are shown as spirals, as in *Laphria*. He apparently assumed that the spermathecae of all Asilidae have the same form, and obviously did not examine the spermathecae of the other two species, which are quite different, as shown below.

There are three spermathecae in all the Asilidae examined, except *Proctacanthus*, *Eccritosia* and *Myaptex*, in which there are only two. Owsley (1946) illustrated the spermathecae of several American species and gave information on the gland and musculature, but did not describe their more detailed structure. All three spermathecae are usually of the same form, but the median spermatheca is of slightly different form in some species (*Polyphonius*). The only exceptions so far found are *Leptogaster* and *Euscelidia* in which the median spermatheca is markedly different from the lateral spermathecae in some species (Figs. 13, 15, 22).

The spermathecae consist of the following parts in most species (Figs. 1–2):

1. A more or less strongly sclerotized tube or capsule, in which the sperm is kept (reservoir). This part is either completely or partly surrounded by a gland with sclerotized, intracellular canaliculi with a terminal reservoir. The canaliculi open into the lumen of the spermatheca. The gland is apparently restricted to the ducts in some genera in which the reservoir is strongly sclerotized (*Leptogaster*, *Apoclea*, *Proctacanthus* and *Efferia*).

2. A duct of varying length, which may also be surrounded by the same gland, usually only the part adjacent to the reservoir, but in some cases far posteriorly. The posterior part of the duct is often more thick-walled or striated.

3. A more or less sclerotized, generally funnel-shaped valve which opens into a differentia-ted part of the ducts. This may be striated or strongly sclerotized, with outer processes, and sometimes with a basal sclerotization. A layer of muscles extends from above the valve to the posterior end of the duct. The muscle fibres are either attached to the outer processes of the tube or extend to its posterior end. The ducts either end separately on the membrane between the arms of the furca into the vagina or open into a common duct of varying length and structure; this may be very short, or wide and crinkled (*Apoclea*), or very long (*Laphystia*).

4. There are two accessory glands which open posterior to the spermathecae into the vagina. The cells of these glands also have intracellular canaliculi. They are of varying length and width; in *Leptogaster* they are very long and narrow and have a basal sclerotization.

The proportions of the parts vary markedly. The ducts may be very short (some species of *Apoclea* and *Polyphonius*) or extremely long and of different structure in their different parts. Some of the differentiations described above (valve, etc.) may be absent or not recognizable in preparations from dry material.

The basal, sclerotized tube, with the valve and its muscular cover, is apparently an ejection

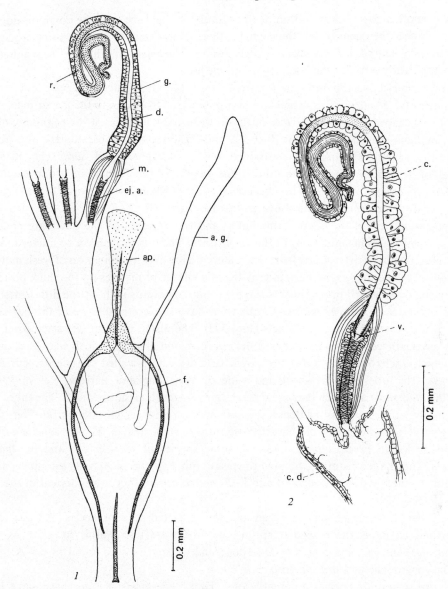

Figs. 1–2: *Dysmachus verticillatus.* spermatheca; 2. spermatheca, details

a.g.—accessory gland; ap.—apodeme of furca;
c.—canaliculi of gland; c.d.—common duct;
d.—duct of spermatheca; ej.a.—ejection apparatus;
f.—furca; g.—gland; m.—muscles; r.—reservoir
v.—valve

apparatus or sperm pump, as already suggested by Snodgrass (1935), and the amount of sperm ejected during oviposition is apparently regulated by the valve at the entrance to the tube. The valve has a distinct, more or less complicated form if it is strongly sclerotized, but it usually contains a funnel-shaped inner part or a number of posteriorly directed spines.

There is another basal sclerotization at the opening of the tube into the common duct or near it, or near the opening into the vagina if there is no common duct (in some species of *Heteropogon, Anarolius, Pycnopogon, Holopogon* and *Stichopogon*). This may bear tubercles which apparently serve for the insertion of muscle fibres.

The valve and ejection apparatus are apparently absent in most genera of Laphriinae, except *Trichardis* and *Hoplistomerus*, which are here considered to belong to the Laphriinae. They have a ring-shaped or cylindrical sclerotization with internal spines at the beginning of the posterior, bulb-shaped part of the ducts. A valve and ejection apparatus are also absent in the five genera of Laphystiini examined and in *Dasypogon* and *Saropogon*, in some species of *Promachus* and in the other three genera of the *Promachus* group examined (*Apoclea, Philodicus, Alcimus*). The absence of these structures may prove to be a group character if further genera of the above groups are examined.

The posterior part of the ducts forms bulbs, and the openings of the ducts are plicated, with 4–5 peripheral invaginations in the Laphriinae. There is probably a sphincter muscle, but this could not be determined from dry material. However, examination of fresh material of *Promachus* and *Saropogon* showed that there is a layer of muscles on the thick posterior part of the ducts which probably acts as an ejection apparatus without the differentiations in the other genera. Some species of *Promachus* have a sclerotized valve at the beginning of the wide posterior part of the ducts (Fig. 395). The distal, more or less sclerotized part of the spermathecae (reservoir) varies widely in form and arrangement. It may be a simple tube which is sclerotized to a varying extent or a wide, thin-walled sac (*Eccoptocus, Pamponerus*). The tubes may be short and wide or very long and narrow, e.g. in *Perasis* and *Rhipidocephala*, in which they are several times longer than the abdomen. The tubes may form an irregular, dense coil, a dense, regular spiral, or a loose, irregular spiral. They form strongly sclerotized capsules of characteristic form in some genera. The capsules are ovoid or cup-shaped in *Proctacanthus*, ovoid in some species of *Habropogon* and in *Apoclea conicera*. They have a recurved, pointed process in other species of *Apoclea* examined, and a straight, pointed end in *Polyphonius* and in *Eremisca osiris*. They are spherical in the new genus A (Fig. 329).

These different forms of spermathecae are not restricted to subfamilies. All species of Laphriinae examined have regular spirals, but there are also spirals in some Asilinae (*Promachus*) and in some genera of Dasypogoninae. Capsules were found in the Asilinae, the Dasypogoninae and in *Leptogaster*.

There are sometimes special differentiations. Thus, the spirals of *Saropogon* have long, spine-shaped processes into the lumen of the tubes from the opening of the canaliculi of the gland. The spermathecae of *Dasypogon* do not form spirals, but have the same internal spines (Figs. 184–186). Similar internal spines in the tubes of the spirals were found also in two species of *Atomosia* (Laphriinae, Fig. 39).

The furca is considered as the modified sternite 9. It consists of a frame with a membrane between the lateral arms in which the ducts, or the common duct, open. It varies widely in form. It is more or less U- or V-shaped in the Laphriinae and Dasypogoninae, but the width of the frame and its form vary widely. It has a short, wide distal apodeme in some species of the above subfamilies; it consists of only two lateral arms connected by a membrane in *Leptogaster* and *Dioctria*.

The furca of the Asilinae differs markedly. It consists of a usually long, slender frame, with a distal apodeme of varying form and length and a posterior median sclerite between the posterior end of the lateral arms which is also of distinctive form in some species. It is Y-shaped in some species of *Proctacanthus* and partly fused with the triangular sclerite beneath the cerci, so that it connects the furca with this sclerite (Fig. 350). The connection is membranous in species in which the sclerite is rod-shaped or absent. The furca is violin-shaped in some species of *Promachus*. It is completely sclerotized in some genera, but is sclerotized only in its distal part with the apodeme and in the posterior part of the lateral arms in other genera (*Neomochtherus*, *Cerdistus* and others). There are plate-shaped sclerotizations in a few genera (*Eremisca*, *Proctacanthus*). Strong muscles are attached to the distal apodeme, which may have a dorsal ridge, and to the posterior median sclerite.

Aedeagus

The only detailed description of the aedeagus, its structure and function, in three species of Asilidae, was given by Reichardt (1929); however, this description contains a number of wrong interpretations.

ASILINAE

The aedeagus (Fig. 3) consists of an outer sheath, within which the aedeagus proper or aedeagus pump is situated. The aedeagus pump is firmly fused with the sheath in its distal

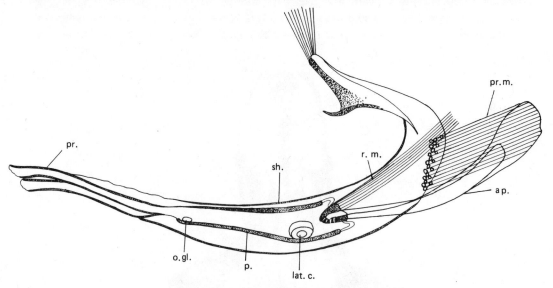

Fig. 3: Aedeagus (Asilini, diagrammatic)

ap.—apodeme; lat.c.—lateral connection; o.gl.—opening of gland;
p.—pump; pr.—prongs; pr.m.—protractor muscles of apodeme;
r.m.—retractor muscle of apodeme; sh.—sheath

part, but free at the base. It is connected at the base with the sheath by a ventral, disc-shaped connection in species in which the sheath is narrow in the apical part and wide only at the base (*Machimus*). It is fused ventrally at the base in a wider area in the *Apoclea* group. It has two cylindrical lateral connections in some species in which the sheath is wide (*Neomochtherus, Cerdistus, Eremisca*), and a single ventral cylindrical connection in *Philonicus albiceps*. In some genera (*Pamponerus*, new genus A) the connection with the sheath consists of two oblique ridges.

In many species the aedeagus pump is divided into three terminal, closed tubes (*Promachus*), or half-tubes ('prongs'). There are only two tubes or prongs in some species, the median prong being reduced or absent. There are only two tubes to near the base in *Proctacanthus, Eccritosia* and *Myaptex*. Martin (1968) states that there are also only two tubes on the aedeagus in *Lecania* and *Proctacanthella*. There are only two tubes in *Philodicus ponticus*, but there are three in *Philodicus spectabilis*.

The sheath may be very narrow and closely adjacent to the aedeagus pump (*Machimus* and related genera), or it may be wider and not so closely adjacent to the aedeagus pump (*Neomochtherus, Cerdistus* and other genera). The aedeagus pump of *Neomochtherus* has a single opening, and the prongs are rudimentary and not functional. The sheath shows various differentiations at the distal end, denticles, bulges, leaf-shaped processes, etc., in some species, and its form and differentiation are different in practically every species examined.

The aedeagus pump is membranous at the base, and articulated with an apodeme of varying form on which protractor and retractor muscles are inserted. These muscles originate on the inside of the sheath at or near its basal margin (Figs. 4–5). The membranous base of the aedeagus pump permits forward and backward movement of the apodeme, which acts like the piston of a syringe. The ductus ejaculatorius opens dorsal to the apodeme into the lumen of the pump, not 'through' the apodeme as stated by Reichardt. This is obviously a

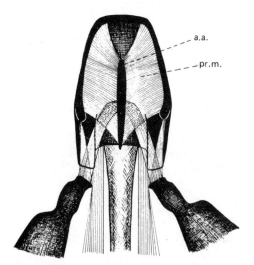

Fig. 4: *Machimus setibarbus*, base of aedeagus, dorsal
a.a.—apodeme of aedeagus; pr. m.—protractor muscles of apodeme

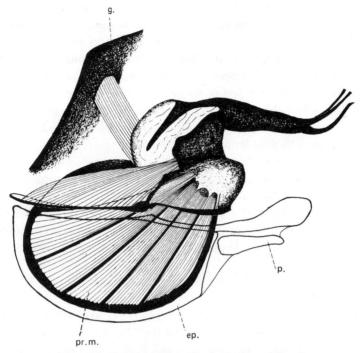

Fig. 5: *Laphria dizonias*, aedeagus, lateral

ep.—epandrium; g.—gonopod; p.—proctiger;
pr.m.—protractor muscles of apodeme

sperm pump, as was already suggested by Wesché (1906). The pump of many species has two rounded openings on the ventral side in which a gland opens (Fig. 391).

The apodeme is a vertical plate of varying form, but it changes markedly in size during the life of the individual by deposition of skeletal material at its free end. It narrows into a 'head' on which two groups of sensillae are situated, the number and arrangement of which differ in the different species. These were considered to be glands by Reichardt.

The muscles mentioned obviously cannot change the position of the pump inside the sheath, as Reichardt assumed, as it is fused with the sheath distally and by the cylindrical connections at the base, and because the muscles originate inside the sheath. Their function is evidently to move the apodeme.

The aedeagus is flexibly connected with the gonopods by two dorsal posterior apodemes at the base of the sheath. These apodemes may be rod-shaped and recurved (*Promachus* group), or of complicated form. They are replaced in a species of *Ommatius* by long tendons which are partly sclerotized. Muscles inserted on these apodemes change the position of the aedeagus.

The proportions of the parts of the aedeagus pump vary markedly in the different genera. The pump is nearly as long as the sheath (*Machimus* group) or much shorter (*Neomochtherus*, *Polyphonius*, *Promachus*). The prongs or tubes may be long or short; they may be of the same length, or the median prong may be shorter or curved across the lateral prongs, very much longer (*Rhadiurgus*) or very much shorter (*Machimus corsicus*) than the lateral

9

prongs. The distal end may show differentiations, e.g. in *Machimus chrysitis*, which induced Hull (1962) to state that 'the aedeagus of the genus *Machimus* is five-pronged'. In fact, *M. chrysitis* has not five but seven prongs, and two further triangular processes at the base of the prongs, but only three prongs are functional, i.e. connected with the lumen of the pump. All other species of *Machimus* examined have only three prongs. In some species the prongs apparently enter the ducts of the spermathecae during copulation. This is supported by the position of the openings of the ducts of the spermathecae in *Rhadiurgus*, in which the opening of the median duct is situated much further anteriorly, corresponding to the ending of the long median prong of the aedeagus.

There are marked differences in the aedeagus of the various species of *Promachus* examined. The Palaearctic species have a short aedeagus pump with three terminal tubes and a long, slender apodeme. The tubes are short cups with internal denticles in *P. griseiventris*, but they are longer than the basal part of the aedeagus in other species and have different differentiations at the end. There are completely different types of endings in some of the Ethiopian species of *Promachus* examined, some of which have an aedeagus with a single opening (Fig. 394).

The aedeagus of *Polyphonius* and of the new genus A resembles that of *Neomochtherus*, but the prongs have been entirely lost.

The aedeagus of species of *Apoclea* and *Philodicus* shows remarkable differentiations of the apical part. The basal part is slender and wide at the base, but the apical part is articulated by a membrane with the basal part and is folded back upon the basal part at rest. The apical tubes are wide, of varying form, and covered with denticles to a varying extent. The lumen of the pump continues in a narrow tube through the apical tubes and ends in a sclerotized cup. The median tube differs in form from the lateral tubes in some species. The aedeagus has a specific form in each of the ten species of *Apoclea* examined. The membranous articulation of the apical part is present in eight of the species of *Apoclea* and in the two species of *Philodicus*, but it is absent in two species of *Apoclea* which have tubular endings of the aedeagus without a membranous articulation. A similar aedeagus with long, tubular endings and without articulation is also present in a species of *Alcimus*. The arrangement is similar in *Philodicus ponticus*, but the median tube is rudimentary, not functional (there is no inner tube), and there is a large process of complicated form on the dorsal side of the apical part. The aedeagus of *spectabilis* has three long, tapering tubes, curled at the end and covered with granules, and the apical part is articulated and folded back on the basal part, as in *P. ponticus*.

According to the illustrations given by Martin (1968) and Artigas (1971) there are further variations. There is apparently a tendency to reduction in some Asilinae from an aedeagus with three prongs to one with short and narrow, but functional prongs, then to an aedeagus with rudimentary, not functional prongs, and finally to a total loss of prongs. This simple aedeagus with a single opening and without prongs would, in this case, not be primitive but secondary. On the other hand, there is a tendency to differentiation, which has resulted in the complicated structures in *Apoclea* and *Philodicus* and in the new genus B described below.

LAPHRIINAE

The aedeagus differs markedly from that of the Asilinae. The sheath is wide, relatively short, and the pump ends in three closed tubes. The sheath forms large apical bulges in its basal part, so that the inner structure of the aedeagus is difficult to recognize without removal of the bulges. The protractor muscles of the apodeme originate in these bulges (Fig. 5). The lumen of the pump contains posteriorly directed spines in some species of *Laphria*. It continues in a narrow tube, which curves ventrally at nearly a right angle behind the sheath in some species, and opens in a wide pump chamber with which the head of the apodeme is articulated. The pump chamber is thus markedly ventrally displaced. The apodeme is very wide and short and apparently curved towards the epandrium, to judge from the form of its inner cavity. The form of the aedeagus and the length of the tubes vary considerably in the different genera. The tubes are relatively short in *Laphria* and *Ctenota*, long in *Andrenosoma*, and extremely long and thin in *Pogonosoma*; the tubes continue inside the sheath in some genera, and in *Andrenosoma* the sheath is divided so that each tube has a separate sheath.

LEPTOGASTRINAE

The aedeagus of the species examined is a simple tube. It is very long, strongly curved, and tapers to a fine point in *Leptogaster cylindrica*; shorter, less curved and wider in other species. There are two long, nearly rectangular basal plates, on which the protractor muscles of the apodeme originate, as in the Dasypogoninae. The apodeme is large, rounded-triangular.

Martin (1968) gives drawings of complicated aedeagi of some Leptogastrinae in addition to simple aedeagi, but it is not clear from the drawings whether these structures are situated on the sheath or whether there is a divided lumen. He states that the aedeagus of *Beamero-myia* is divided apically into two tubes.

DASYPOGONINAE

Most Dasypogoninae have a simple, conical aedeagus with a wide pump chamber and a large apodeme, but the pump chamber is not ventrally displaced as in the Laphriinae. There are differentiations on the distal part of the sheath, which may bear spines, denticles, horns, bulges or apical flanges. The aedeagus may be very short and wide or slender and conical. It may be straight, curved, or S-shaped. The sheath bears two long apical processes parallel to the apical part of the aedeagus in some species, so that it appears 3-pronged (*Jothopogon*, *Lasiopogon*, *Perasis*).

The only aedeagus with three functional prongs was found in species of *Holcocephala* (Fig. 145) of the tribe Damalini. An aedeagus with three short tubes was also illustrated by Martin (1968) for *Trigonomima* (Damalini).

There are marked differences in the form of the posterior apodemes and in the arrangement of the muscles from that in the Laphriinae and the Asilinae. The posterior apodemes extend obliquely posteriorly and are not curved dorsally as in the Asilinae. They are connected with the inner dorsal margin of the gonopods. The protractor muscles of the apodeme originate on two lateral basal plates which are connected to the membranous pump chamber, not on the base of the sheath as in the Asilinae (Figs. 6–7). The form of these sclerites varies widely in the same genus (e.g. in *Rhipidocephala obscura* and in *Rhipidocephala* sp. (Figs. 167, 169)). The origin of the protractor muscles of the apodeme on separate

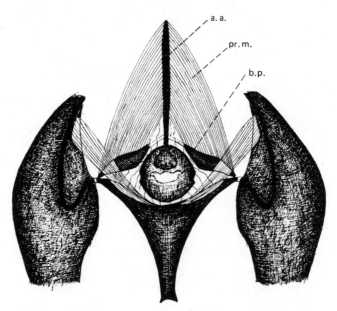

Fig. 6: *Pycnopogon mixtus*, aedeagus, ventral

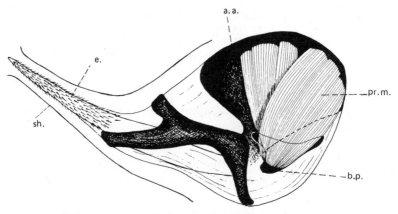

Fig. 7: *Stenopogon* sp., aedeagus, inner part with endoaedeagus

a.a.—apodeme of aedeagus; b.p.—basal plate;
e.—endoaedeagus;   pr.m.—protractor muscles of
apodeme; sh.—sheath of aedeagus

sclerites in the Dasypogoninae and Leptogastrinae, in contrast to the condition in the other two subfamilies, suggests a closer relationship between these two subfamilies.

A peculiar differentiation is present in a number of genera. There is a well defined, pointed tube inside the lumen of the aedeagus pump which widens at the base and is connected with the head of the apodeme. The walls of the tube are perforated with pores, and the tube is covered with posteriorly directed spines in some species, with granules or platelets in others. This is perhaps a continuation of the ductus ejaculatorius and, in this case, the 'true aedeagus' or an 'endoaedeagus'. It may project from the apical opening or be situated inside the pump, its position depending on that of the apodeme (Figs. 7, 59, 64, 73, 76, 77, 94, 100). This structure is very delicate in some cases and may not always be recognizable in material treated with KOH, or may be lost during maceration. It was found only in a few species of Asilinae (*Eremisca*, *Eccoptocus*, *Machimus*), in which it is even more delicate and without spines or granules.

The apodeme is a vertical plate in most genera, but it has a transverse plate at the free end in *Stichopogon*, so that it appears T-shaped in dorsal view. The head of the apodeme has a pointed process into the lumen of the endoaedeagus in some species.

The aedeagus shows distinct specific differences in genera of which several species were examined, e.g. in *Saropogon*.

The position in Laphystiini is rather puzzling. The aedeagus of *Hoplistomerus* and *Trichardis* is divided into three tubes to near the base of the free aedeagus, as in the Laphriinae. The spermathecae of these two genera form large, sclerotized spirals, also as in the Laphriinae.

Scytomedes has a simple conical aedeagus resembling that of other genera of Dasypogoninae. The aedeagus of *Laphystia* is an extremely long, thin, tapering tube with a wide base. *Perasis* has a relatively long, slightly S-curved aedeagus, which is divided internally into two tubes. The aedeagus of *Triclis* is similar, but is a simple tube.

The spermathecae of *Perasis*, *Triclis* and *Scytomedes* are simple, extremely long, thin tubes which do not form a spiral or form a loose spiral. Those of *Laphystia* are similar, but the ducts unite in a very long common duct.

The close relationship of the Laphystiini to the Laphriinae has been stressed repeatedly. Oldroyd (1963) placed them in his tribe Laphriini. The structure of the aedeagus and spermathecae of *Hoplistomerus* and *Trichardis* clearly indicates that they belong to the Laphriinae, and this would also apply to other genera of Laphystiini with a similar aedeagus and similar spermathecae.

The tribe Laphystiini (Prytaniinae of Hermann and Engel, Hoplistomerini of Karl) has been based mainly on the reduced number of abdominal segments. However, the posterior abdominal segments are also reduced to a varying extent in other groups of Asilidae, and the Laphystiini are probably not a homogeneous group according to the structure of the aedeagus and spermathecae.

Gonopods, Proctiger and Ovipositor

There are other parts of the genitalia which have usually not been described in sufficient detail. This applies particularly to the gonopods and the form and chaetotaxy of the dististylus, which show generic characters in some genera, but specific characters in nearly every species examined. The dististyli of certain genera of Asilinae enclose the aedeagus laterally and enter the vagina together with the aedeagus during copulation, as described by Reichardt. They are usually covered with proximally directed spines and setae, which probably hold the aedeagus in the vagina during copulation. The form of the dististylus and its complement of spines and setae should be described and illustrated in detail for every species. The gonopods may also bear other processes of distinct form or groups of spines. However, the form of the processes may vary markedly in the same species, e.g. in *Anisopogon hermanni* (Figs. 103–105).

The gonopods have a proximal apodeme which varies markedly in form. It may be wide and fan-shaped (*Philonicus, Cerdistus syriacus*), T-shaped or rod-shaped (*Promachus, Apoclea* group).

The male proctiger may differ in form and sometimes bears characteristic processes (*Cophinopoda, Efferia, Saropogon*) or characteristic armatures of spines, e.g. in species of *Stichopogon* (Figs. 135–139) and in *Lasiopogon* (Fig. 142).

Sternite 8 of the female is of characteristic form in many species. It forms the ventral part of the ovipositor in the Asilinae and is folded in half, so that its exact form is not recognizable in pinned specimens. It is spread and mounted dorso-ventrally during the preparation of the spermathecae, which are attached to it. It sometimes gives good characters for the distinction of closely related species in which external characters are very similar, e.g. some species of *Machimus* (Figs. 281–282) and the new genus B (Figs. 335–336).

It thus seems necessary to include a detailed description of the genitalia of both sexes in the description of new species, and a redescription of the genitalia of known species would be of great help in the identification of genera in which species have been described on the basis of external characters alone. The description of the genitalia should be based not only on their appearance in pinned specimens, but on separately mounted parts after treatment with KOH. Even the form of the epandrium is usually not described in sufficient detail. The posterior processes may be curved inwards and covered with setae, and their exact form and chaetotaxy are recognizable only in separately mounted parts.

SYSTEMATIC PART

LEPTOGASTRINAE

Leptogaster cylindrica, gracilis, guttiventris and three undescribed species (Figs. 8–15).

THE LATERAL spermathecae are distinctly different in form from the median spermathecae in two of the species examined.

cylindrica. The distal part of the spermathecae forms a strongly sclerotized, thick-walled tube with rounded end. It continues in a short, wide part with thick, twisted processes with a thin end, which are apparently modified canaliculi of the gland cells. This part continues in a straight, tubular part with irregular discs at both ends. The three ducts then unite after a curved, striated part into the long, strongly sclerotized common duct. The furca consists of two lateral bars connected by a membrane. The bars are covered with short, regularly arranged denticles. Fresh material would be necessary to determine the function of the thick-walled, straight tube, which resembles the ejection apparatus in other genera.
The aedeagus is a simple, curved tube, tapering to a fine point. Apodeme large, rounded. Lateral basal plates large, rectangular.

Species no. 1. The spermathecae resemble those of *cylindrica* in general arrangement, but the ducts from the reservoir to the straight part with terminal discs are narrower and much longer and covered with long, thin canaliculi. The discs are much larger and more plate-shaped. The common duct is twice or three times as long as in *cylindrica* and curled in its distal part. Furca as in *cylindrica*, but the denticles on the bars are longer and less regularly arranged.
Aedeagus as in *cylindrica*, but much longer and thinner.

gracilis. The reservoir of the spermathecae forms a densely coiled, thick-walled tube with rounded end, continuing into a thick duct with thick, twisted canaliculi in the distal part. The ducts unite in a thick, sclerotized, relatively short common duct. The median spermatheca is thinner and with more numerous and densely arranged canaliculi on a shorter area. Furca with two separate bars, which are pointed at both ends.
Aedeagus with a short, tapering, narrow tube and a wide, rounded basal sheath. Basal plates very long, nearly rectangular.

guttiventris. Spermathecae with tubular reservoirs as in *cylindrica*, continuing into a narrow duct with long canaliculi, and into a long, thin, thick-walled part, but terminal discs are absent. The common duct is shorter than in *cylindrica*. The lateral bars of the furca are pointed at both ends.
Aedeagus as in *cylindrica*.

Species no. 2. The lateral spermathecae have a tubular reservoir like the preceding species but the median spermatheca is a wide, ovoid-conical capsule. Ducts long, narrow, with canaliculi which begin at a distance from the capsule. Posterior part of the ducts without

15

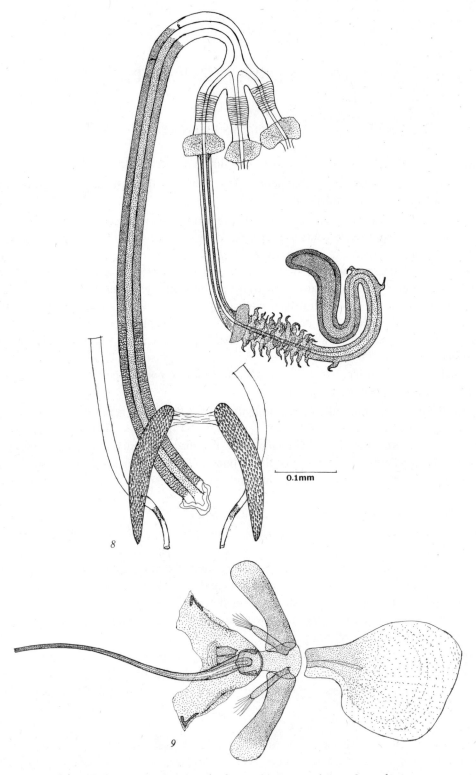

Figs. 8–9: *Leptogaster cylindrica*. 8. spermatheca; 9. aedeagus

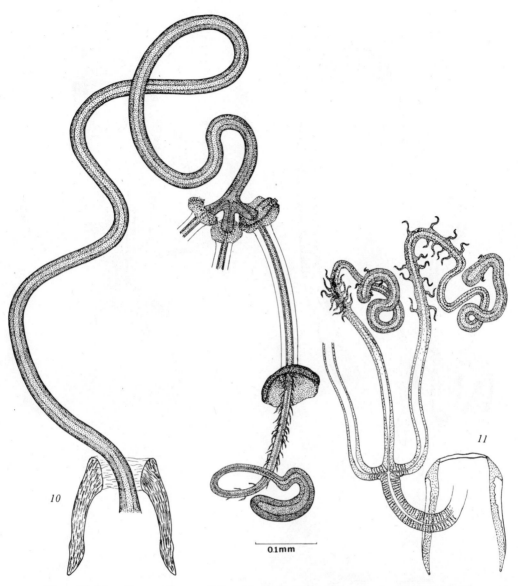

Figs. 10–11: Spermatheca of *Leptogaster*. 10. species no. 1; 11. *L. gracilis*

canaliculi and striated at the base. The ducts unite in a sclerotized ring. Common duct either absent or short and membranous. Lateral bars of furca wider anteriorly and closer together than in the other species.

Aedeagus short, tapering to a fine point.

Species no. 3. The median spermatheca is a strongly sclerotized tube with rounded end. It continues in a slightly wider part with canaliculi, and then in a straight, striated part with a distal and a proximal extension with lateral processes. The two lateral spermathecae form wide, ovoid capsules with a terminal, recurved, narrow tube. Common duct short, strongly sclerotized, striated. Lateral bars of furca pointed at both ends.

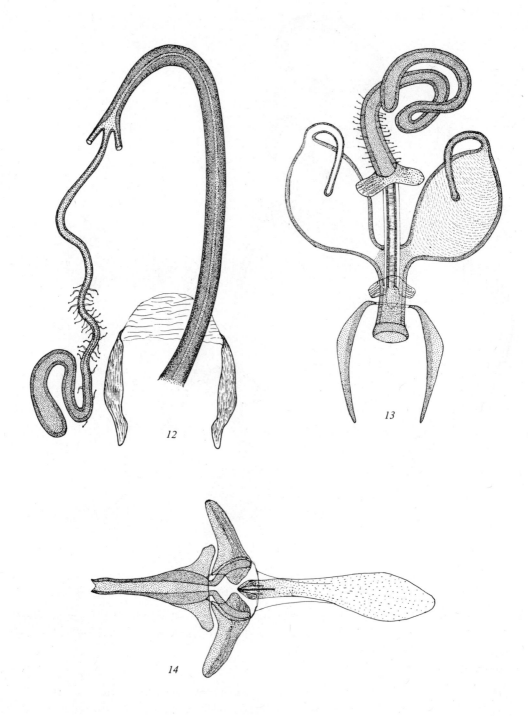

Figs. 12–14: *Leptogaster.* 12. *L. guttiventris*, spermatheca;
13. species no. 2, spermatheca; species no. 3, aedeagus

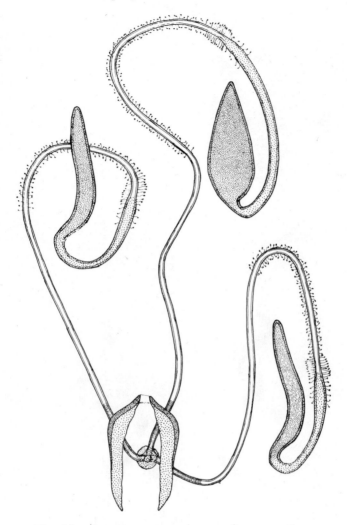

Fig. 15: *Leptogaster*, species no. 3, spermatheca

Aedeagus simple, straight, slightly tapering, much shorter and wider than in *cylindrica*.

Euscelidia artaphernes, *bishariensis* and an unidentified species from South Africa (Figs. 16–22).

The male genitalia of all three species examined are markedly modified. The gonopods are strongly reduced and displaced to near the apex of the hypandrium, so that the epandrium borders on the hypandrium.

artaphernes. Aedeagus very long, narrow, with a narrow inner tube. Sheath narrow, curved at an angle at the pointed apex. Base of sheath conical, curved, with broad dorsal apodemes. Pump chamber small, apodeme short, with wide head and without neck. Basal plates large, deeply bifid.

Gonopods very small, with an oblong-oval dististylus and a similar apical process. Hypandrium large, truncate-triangular, with an apical frame which encloses a membranous area.

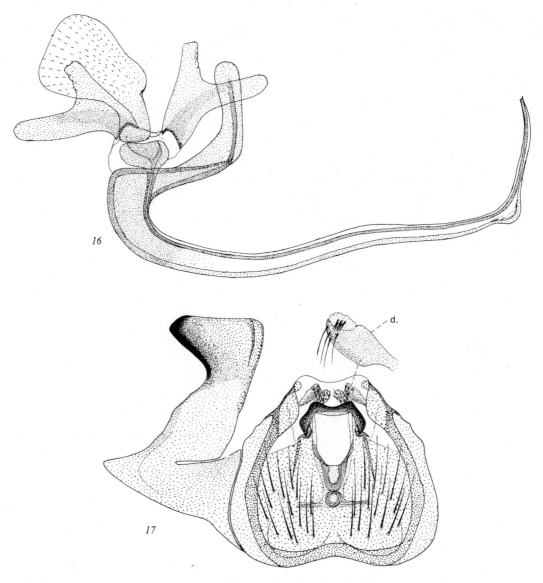

Figs. 16–17: *Euscelidia ataphernes*. 16. aedeagus;
17. epandrium, hypandrium, gonopods and dististylus (d.)

bishariensis. Aedeagus short, broad, with a narrow inner tube. Sheath with a large dorsal apical bulge. The dorsal basal apodemes are fused at the apex and form a triangle. Basal apodemes rectangular, long. Apodeme short.

Gonopods displaced to sides of apex of hypandrium, with a narrow dististylus and a small rounded process, and with two sclerites which end in a curved horn and are connected by a bridge in the middle. Hypandrium large, truncate-triangular, with a broad, parallel-sided median groove at the apex.

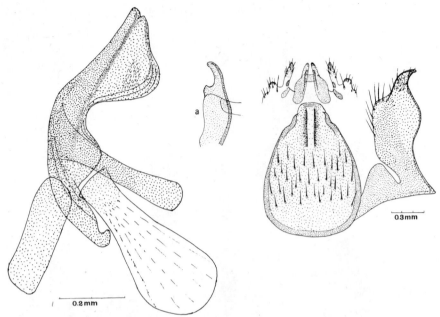

Fig. 18: *Euscelidia bishariensis*, aedeagus, epandrium, hypandrium and gonopods; (a) median process of gonopods, lateral

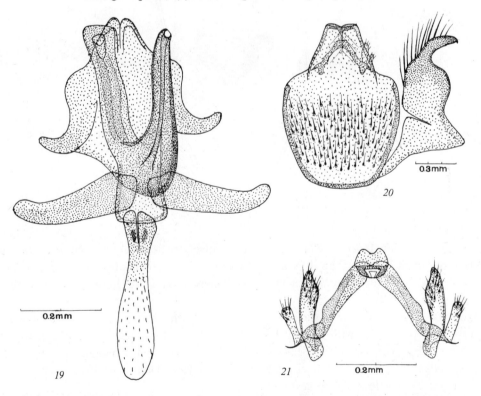

Figs. 19–21: *Euscelidia* sp. (South Africa). 19. aedeagus; 20. epandrium and hypandrium; 21. gonopods with median process

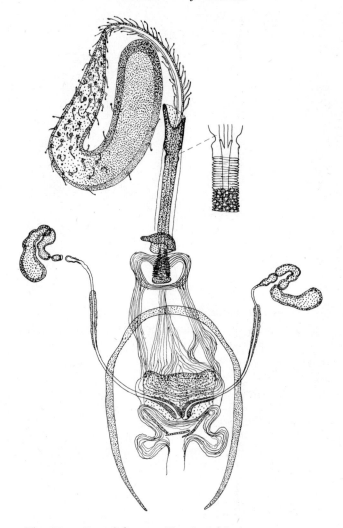

Fig. 22:   *Euscelidia* sp. (South Africa), spermatheca

Species from South Africa. Aedeagus short, relatively wide, slightly S-curved, with an inner tube which is wider than in *artaphernes* and widens proximally. Sheath with a large basal plate with rounded lateral corners and bifid apex. Basal plates long, truncate-triangular, narrower laterally.

Gonopods less reduced than in the other two species, with a spindle-shaped dististylus and a narrow lateral process with setae at the end, situated at the sides of the apex of the hypandrium. There is a V-shaped sclerite at the apex with a rounded plate which apparently corresponds to the sclerite with horns in *bishariensis*. Hypandrium large, broad, with rounded sides and a truncate apical process.

The difference between the median and the lateral spermathecae is even more marked than in some species of *Leptogaster*. The median spermatheca forms a large, strongly sclerotized tube which becomes less sclerotized proximally and tapers into a thin duct. Canaliculi very long, twisted, but not as thick as in *Leptogaster*, present on the proximal part of the reservoir and on the ducts. The duct then passes through a wide, straight, strongly sclerotized tube with reticulate structure and has a valve inside this tube which is thus apparently the ejection apparatus. The tube has a concave sclerotization before its proximal part, which is spindle-shaped and has distinct ridges. It opens in a very wide membranous duct. The lateral spermathecae are very small and open far posterior to the median spermatheca. They form a club-shaped, coiled, sclerotized tube with a few canaliculi and deep constrictions. Ducts narrow, ending close together in a thicker part with ridges. The constrictions are so deep in one spermatheca that the tube appears to be divided into compartments connected by a narrow duct, and it is apparently closed entirely at one point, so that it is doubtful whether it is functional; however, this may be an abnormality. Furca U-shaped, with narrow arms.

LAPHRIINAE

Laphria aurea, dizonias, flava, ephippium, gibbosa (Figs. 23–24)

The spermathecae form large, sclerotized spirals which are surrounded by the gland. They continue in a narrow duct, which is wider and thick-walled in the posterior part, narrows again, and then forms a large, conical bulb. The ducts open separately. Their opening is plicated, with 4–5 peripheral invaginations. There is no recognizable differentiation of the ducts, except sclerotization of their inner walls in some species. The furca is U-shaped, but shows differences in the various species.

Aedeagus.

flava. Sheath slightly wider at the base, tubes about half as long as the basal part, nearly straight. Spines in the distal part of the lumen small.

ephippium, gibbosa. Aedeagus similar to *flava*, but sheath narrower, tubes shorter, about a third or a quarter as long as the basal part.

aurea. Sheath slightly wider at the base, tubes wide, as long as the sheath.

dizonias. Sheath nearly tubular, tubes curved, about half as long as the sheath.

Choerades gilva, ignea, marginata (Fig. 25)

Spermathecae as in *Laphria*. Furca with lateral distal corners.

Aedeagus.

gilva, ignea. Sheath wide in the distal part, with a large ventral bulge, narrow at the base. Tubes short, thin, a third as long as the sheath. Lumen of pump with distinct, posteriorly directed spines.

marginata. Tubes moderately wide, about half as long as the sheath, which is short and tubular.

Maira sp. (New Guinea)

Spermathecae as in *Laphria*. Furca with lateral distal corners.
Aedeagus as in *Laphria*. Tubes nearly as long as sheath, curved at the end.

23

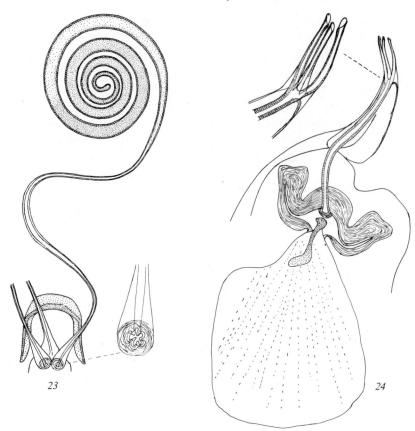

Figs. 23–24: *Laphria flava*. 23. spermatheca; 24. aedeagus

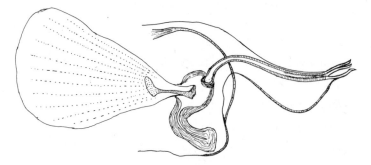

Fig. 25: *Choerades gilva*, aedeagus

Andrenosoma atra (Fig. 26)

The spermathecae form a strongly sclerotized spiral of 2–3 turns. They narrow abruptly into a narrow duct which opens at the side of the spiral or its end. Furca U-shaped, with lateral extensions.

Aedeagus divided completely into three long tubes which taper to a fine end. Sheath divided so that each tube has a separate sheath.

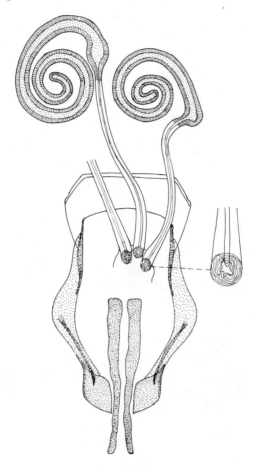

Fig. 26: *Andrenosoma atra*, spermatheca

Pogonosoma maroccanum (Fig. 27)

Tubes in spiral of spermathecae narrow, sclerotized, forming 4–5 turns. Ducts relatively short, their inner wall partly sclerotized. Furca rectangular, very wide anteriorly, with two inner and two narrow posterior processes.

Aedeagus with narrow, curved sheath. Pump divided to near the base into three very long, thin, tapering tubes which are nearly three times as long as the sheath. Head of apodeme small, pointed, with large sensillae.

Ctenota molitrix, Ctenota sp. (Figs. 28–29)

The spermathecae form wide, thin-walled, tubular sacs in a loose spiral. They have 5–8 constrictions in *molitrix*, which are less marked or are absent in *Ctenota* sp. Proximal bulbs little marked. Furca broadly U-shaped, of slightly different form in the two species. Aedeagus with a short, wide sheath and three wide tubes which are nearly as long as the

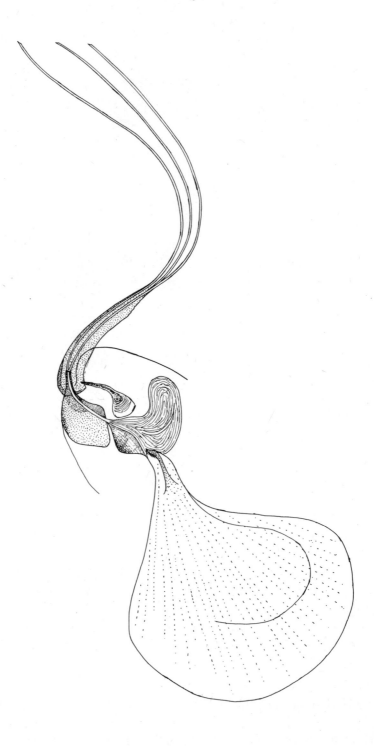

Fig. 27: *Pogonosoma maroccanum*, aedeagus

Fig. 28: *Ctenota molitrix*, spermatheca

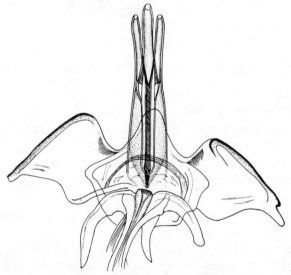

Fig. 29: *Ctenota* sp., aedeagus

sheath. Sheath with three wide, triangular apical processes in *Ctenota* sp., without lateral apical processes in *C. molitrix*.

Processes of gonopods, dististylus, epandrium and proctiger with distinct differences in the two species. Hypandrium small, triangular, without setae.

Lamyra vorax (Fig. 30)

The spermathecae form very wide, thin-walled tubes which form a loose spiral of 1–2 turns. Ducts short, wide, not widening in the posterior half. Furca broadly U-shaped, with a plate-like extension in the curve.

Aedeagus divided into three wide tubes to near the base. Tubes curved at the end. Head of apodeme very wide. Hypandrium triangular, without setae.

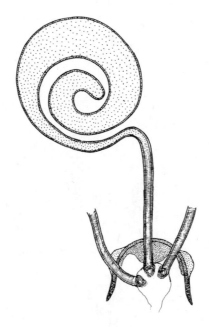

Fig. 30: *Lamyra nobilis*, spermatheca

Stiphrolamyra rubicunda (Figs. 31–32)

The spermathecae form a thick spiral with two turns and rounded end. They are slightly sclerotized. Ducts short, with a proximal sclerotization with tubercles, followed by a wider membranous part and then by a sclerotized ring. The short, membranous ducts behind the ring open in a common opening. Furca U-shaped, with thin arms and a plate-shaped widening in the curve.

Aedeagus with three tapering tubes which are curved at the apex. Basal free part short. Sheath with wide, rounded base. Apodeme very large.

The gonopods resemble those of *Ctenota*. Dististylus with very broad, nearly rectangular base and narrow, black apical part. Apical process parallel-sided, with dorsal point. Hypandrium small, trapezoidal, without setae.

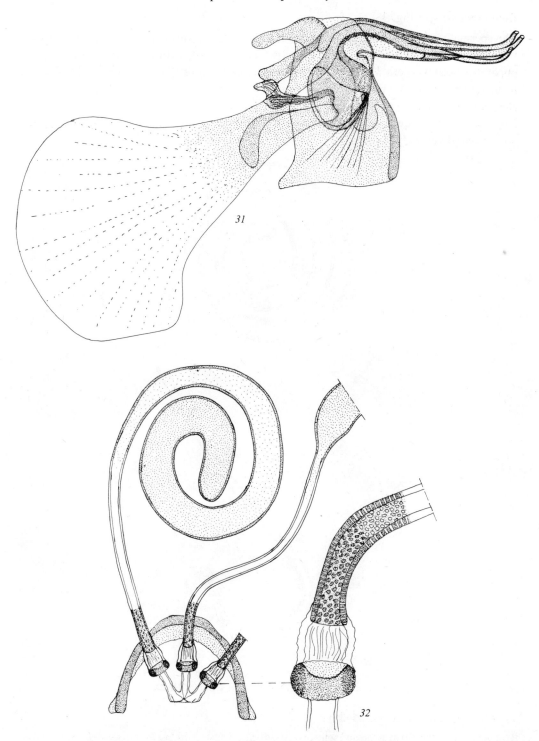

Figs. 31–32: *Stiphrolamyra rubicunda.* 31. aedeagus; 32. spermatheca

Hoplistomerus sp. (Fig. 33)

The spermathecae form large, sclerotized spirals with 2–3 turns and rounded end. Ducts wide, narrower in the middle. Valve cylindrical, sclerotized, with internal spines. Ejection apparatus weakly sclerotized, with an inner sclerotization at the proximal end. The ducts form a long bulb with a plicated opening, as in *Laphria*.

Aedeagus divided into three slightly curved tubes to near the base. Sheath with wide lateral bulges and with large apical processes with rounded end. Apodeme large and wide, but not displaced ventrally.

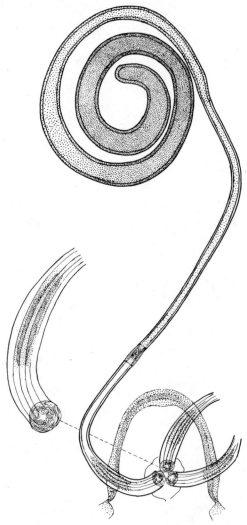

Fig. 33: *Hoplistomerus* sp., spermatheca

Trichardis leucocoma (Figs. 34–35)

The spermathecae form a large, sclerotized spiral with 4–5 turns and rounded end. Outer turn of spiral with reticulate structure. Ducts wide. Valve sclerotized in a narrow ring, with

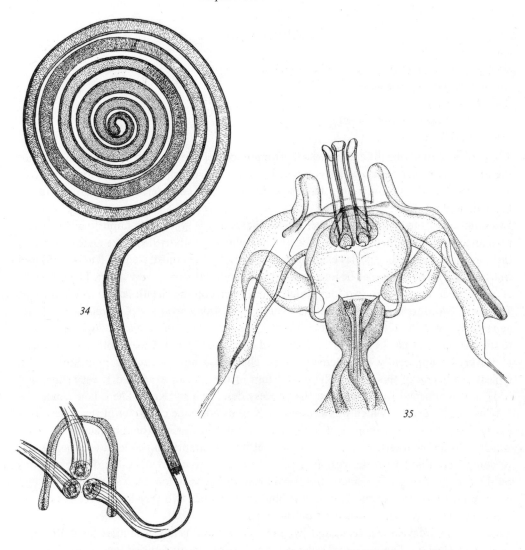

Figs. 34–35: *Trichardis leucocoma*. 34. spermatheca; 35. aedeagus

internal spines. Ejection apparatus with a narrow bulb and plicated opening. Furca U-shaped, narrower than in *Hoplistomerus*.

Aedeagus divided into three tubes to near the base. Sheath with wide lateral bulges and two cylindrical processes with rounded end, parallel to the tubes. Apodeme with large head with two large fields of sensillae. The aedeagus resembles that of *Ctenota*. Hypandrium triangular, with long, pointed apex and a posterior indentation.

Atomosiini

Atomosia melanopogon, puella, sayii, Atomosia sp.
Atoniomyia pinguis, viduata
Aphestia annulipes
Cerotainia albipilosa, macrocera
(Figs. 36–52)

The species of this tribe differ so markedly from other Laphriinae in their small size, habitus, the strongly convex tergite 6 which completely covers the very small genitalia, the sclerotized area behind the hind coxae etc., that some authors have considered their position in the Laphriinae doubtful.

Examination of the genitalia, however, shows that they are typical for Laphriinae. They are permanently rotated through 180°, and the epandrium is undivided. The aedeagus forms three apical tubes in the species examined and in the species illustrated by Karl (1959) and Martin (1968). The tubes may be long or short and the basal part is very short. The aedeagus of *Atoniomyia viduata* has two long, thick apical horns on the sheath lateral to the prongs. There are two dorsal apodemes on the aedeagus of *Atoniomyia* and *Cerotainia* which are apparently movable. Karl (1959) described the male genitalia of *Atomosia dispar* and came to the conclusion that they show 'fundamental differences' from those of other Laphriinae. However, he apparently considered specific characters and secondary modifications as typical characters of the group. The epandrium of the species examined is very short and wide, but of normal form, without the process described by Karl. There is a small dististylus, the absence of which was stressed by Karl as a unique character of the group. It is partly fused with the gonopod in some species but has distinct tonofibrillae at the base which indicate the insertion of a muscle, so that it is apparently movable. It is either absent or completely fused with the gonopod in *Aphestia*. There are remarkable differentiations on the proctiger of some species. *Atoniomyia pinguis* has very long, thick spines at the end of the apical processes of the ventral part, and *Aphestia* has two very long processes on the ventral part and two short processes on the dorsal part.

There is a well developed, separated hypandrium without setae. A similar hypandrium is present in *Ctenota, Lamyra* and *Trichardis*, but it is absent in *Laphria* and related genera. The male genitalia of the Atomosiini are thus reduced in size and simplified, but there are no fundamental differences from those of other Laphriinae, and the structure of the genitalia confirms the position of the Atomosiini in the Laphriinae.

The spermathecae of the species of *Atomosia* examined form a sclerotized spiral, and there is no valve or ejection apparatus as in other Laphriinae. The ducts end separately in plicated openings with invaginations which also closely resemble those of other Laphriinae. The spirals have internal spines at the openings of the canaliculi of the gland cells, resembling those in *Saropogon*. The spines are smaller and less numerous in *Atomosia sayii*. The spermathecae of *Cerotainia* form an oblong, flattened spiral with narrow tubes and without spines. The furca is rectangular, U-shaped in *Atomosia* and *Cerotainia*. *Atoniomyia viduata* has similar spermathecae, without spines and longitudinally striated. The furca has a small, rectangular apical part and wide, long, laterally diverging arms.

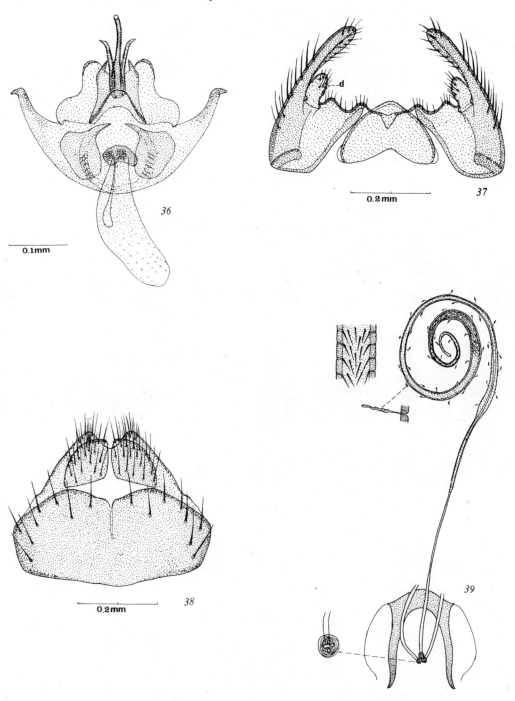

Figs. 36–39: *Atomosia puella*. 36. aedeagus; 37. gonopods, hypandrium and
dististylus (d.); 38. epandrium and proctiger; 39. spermatheca

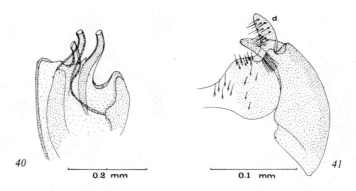

Figs. 40–41: *Atomosia sayii*. 40. aedeagus, lateral;
41. gonopod and dististylus (d.)

Fig. 42: *Cerotainia macrocera*, spermatheca

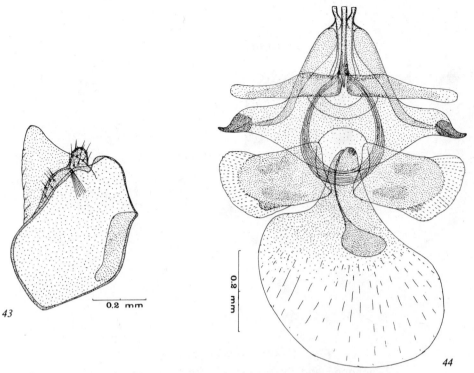

Figs. 43–44: *Cerotainia albipilosa*
43. gonopod and dististylus; 44. aedeagus

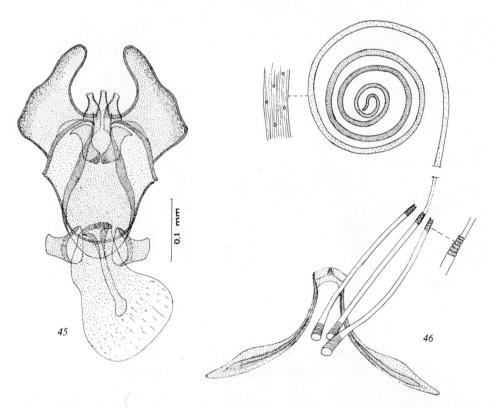

Figs. 45–46: *Atoniomyia viduata*. 45. aedeagus; 46. spermathecae

35

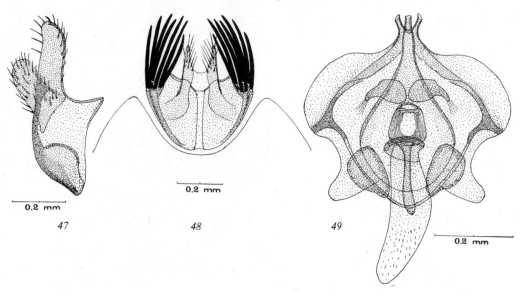

Figs. 47–49:   *Atoniomyia pinguis*
47. gonopod and dististylus; 48. proctiger; 49. aedeagus

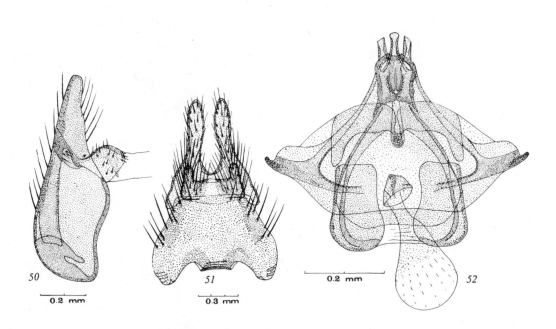

Figs. 50–52:   *Aphestia annulipes.*   50. gonopod;
51. epandrium and proctiger; 52. aedeagus

Laphystiini

Laphystia erberi (Figs. 53–54)

The spermathecae form very long, thin sclerotized tubes which do not form spirals. Ducts narrow, with fine processes in the proximal part. There follows a short part without processes, and the ducts unite in a very long common duct which is striated and crinkled transversely. Furca U-shaped, with narrow, S-shaped arms.

The aedeagus forms a very long, thin, curved tube which tapers to a point, corresponding

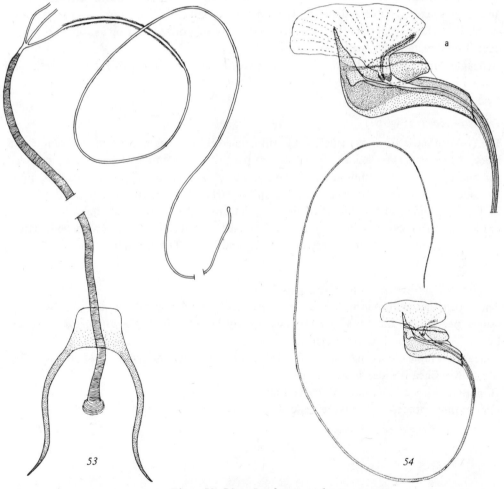

Figs. 53–54: *Laphystia erberi*
53. spermatheca; 54. aedeagus; (a) base of aedeagus, enlarged

to the long common duct of the spermathecae. Sheath short, conical, curved near the base. Apodeme short, very wide.

Scytomedes haemorrhoidalis and an undescribed species (Fig. 55)

The spermathecae form very long, thin, sclerotized tubes which are loosely coiled. Ducts wider, weakly sclerotized towards the base, crinkled in the posterior part, ending in a short common duct, without differentiation. Furca U-shaped, with a wide, rectangular apodeme.

Aedeagus narrowly conical, slender. Sheath nearly tubular anteriorly, with two large, triangular apical processes on the wide part. Apodeme large, rounded. Basal plates large, rectangular. The form of the dististylus and of the apical processes of the sheath of the aedeagus differs distinctly in the two species.

Perasis sp. (Fig. 56)

The spermathecae form extremely long, thin tubes, apparently several times as long as the abdomen, sclerotized in the distal part, ending in a small club. They form a wide, flattened spiral. Posterior part of ducts thick-walled, wider, not sclerotized, ending in a short common duct. Furca rectangular, with narrow arms.

Aedeagus long, slender, slightly S-curved, divided internally into two tubes. Sheath with two long, club-shaped processes ventral to the aedeagus, which are nearly as long as this. Apodeme large, rounded.

Triclis olivaceus (Fig. 57)

The spermathecae form extremely long, thin, sclerotized tubes which taper to a fine end. Ducts wider, thick-walled in the posterior part, then becoming delicate and crinkled, and ending in a short common duct. Valve and ejection apparatus not recognizable. Furca U-shaped, with narrow lateral arms, wide apical part and broad apical apodeme.

Aedeagus simple, long, tubular, S-curved, bifid and with denticles at the end. A sharp, proximally directed denticle near the end on the dorsal side. Sheath with wide lateral bulges. Apodeme short, with very wide neck and head. The genitalia resemble those of *Perasis*.

Glyphotriclis ornata

The spermathecae form long, sclerotized tubes in an irregular spiral. Ducts wider and thick-walled posteriorly and with fine processes. They become thinner again before the end. Furca U-shaped, with narrow arms.

Aedeagus simple, long, narrow, slightly S-curved, nearly uniformly wide in the apical two-thirds. Sheath wide, apodeme large.

Gonopods with a long, narrow, curved lateral process with a spine at the end and with a thicker inner process with pointed end. Dististylus thick, with rounded end.

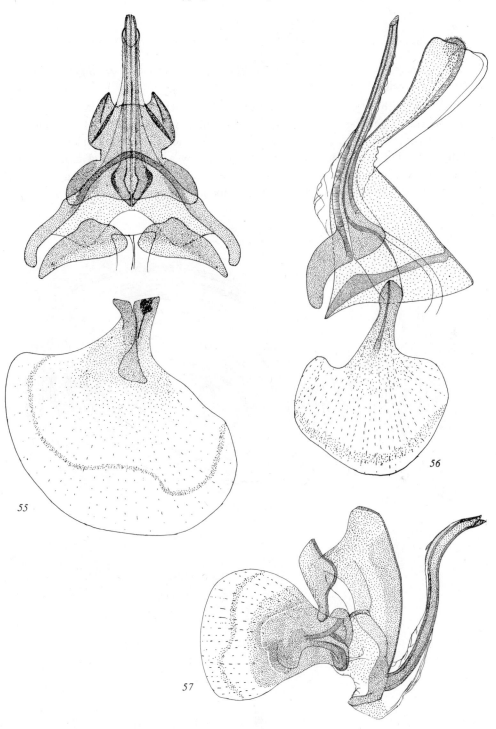

Figs. 55–57: Aedeagus

55. *Scytomedes haemorrhoidalis*; 56. *Perasis* sp.; 57. *Triclis olivaceus*

Stenopogonini

Stenopogon heteroneurus, junceus, laevigatus-milvoides, sabaudus, xanthotrichus and four undescribed species (Figs. 58–62)

Spermathecae with very long, thin, sclerotized tubes which form a regular spiral (7–8 turns in *junceus* and some other species, only 3–4 turns in *laevigatus-milvoides*). The ducts widen towards the base and are covered by the gland to near the valve. This is strongly sclerotized, broadly conical. Ejection apparatus short, covered by a thick layer of muscles from far above the valve. The ducts continue in a wide, crinkled part with circular musculature. Furca U-shaped or V-shaped, with a short apodeme in some species. The spermathecae

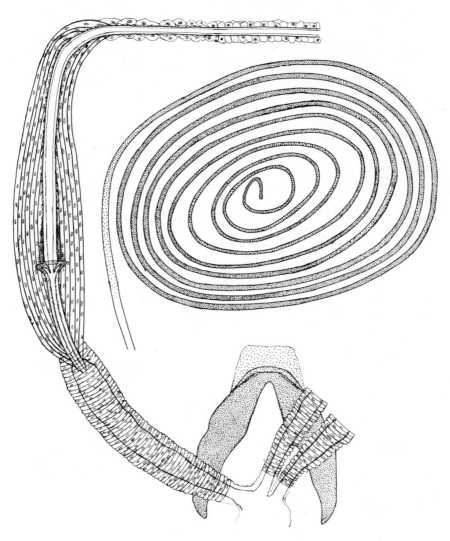

Fig. 58: *Stenopogon* species no. 1, spermatheca, fresh material

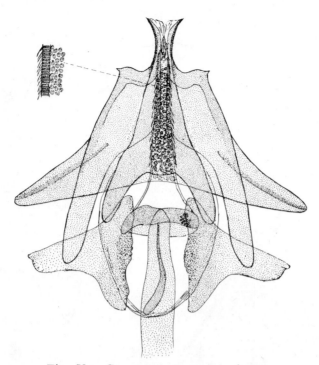

Fig. 59: *Stenopogon junceus*, aedeagus

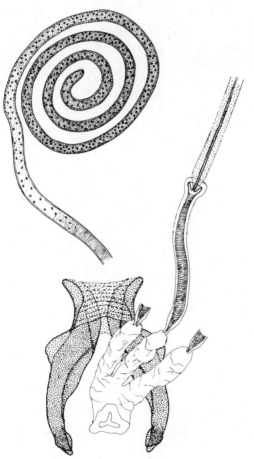

Fig. 60: *Stenopogon laevigatus-milvoides*, spermatheca

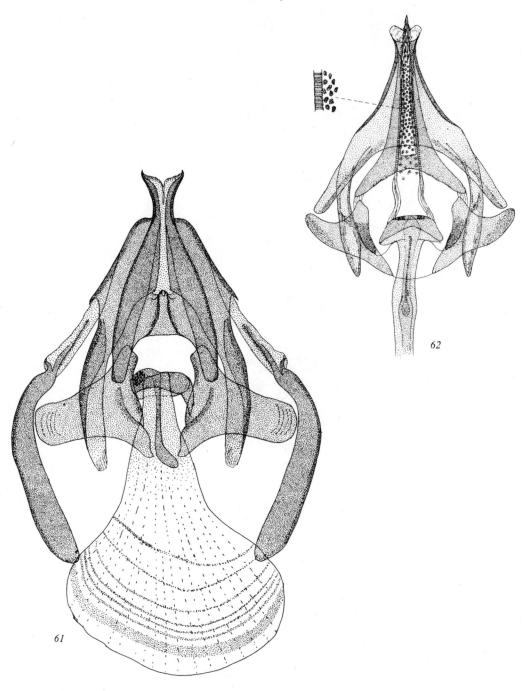

Figs. 61–62: Aedeagus of *Stenopogon*
61. *S. laevigatus-milvoides*; 62. species no. 2

show only minor differences in most species examined, except *laevigatus-milvoides*, in which the spiral is much shorter and thicker and the ejection apparatus has internal spines in the posterior part. The furca of this species is of a more complicated form.

Aedeagus conical, with funnel-shaped end. Head of apodeme T-shaped in dorsal view, wide. The sheath has either pointed lateral apical corners (*junceus*), or rounded apical bulges, or bulges are absent. Endoaedeagus distinct, with spines and platelets.

Gonopods with an apical, dorsally curved hook, and 2–3 apical processes, one of them plate-shaped in some species. Dististylus narrow, with hooked end in some species, of characteristic form in every species examined. Hypandrium of distinct form and with characteristic chaetotaxy in many species, truncate-triangular, with lateral processes, triangular, with bifid apex in *heteroneurus*.

Galactopogon sp. (Fig. 63)

Spermathecae with thin, sclerotized tubes forming a spiral. Ducts very long, thin-walled. Valve distinct, conical, weakly sclerotized. Ejection apparatus long, delicate. Furca broadly U-shaped, with wide arms, a median ridge and a narrow apodeme.

Aedeagus simple, conical, with slightly funnel-shaped end. Sheath with two rounded- or pointed-triangular lateral processes in the apical part. Apodeme large, with T-shaped head in dorsal view.

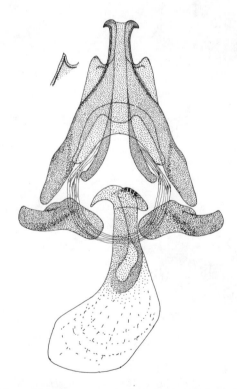

Fig. 63: *Galactopogon* sp. (near *hispidus*), aedeagus

Holopogon dimidiatus, fumipennis, nigripennis and an undescribed species (Figs. 64–71)

Some species of *Holopogon* have been described mainly according to colour characters (wing pattern, pattern of mesonotum, colour of setae). These are difficult to identify, particularly the females, as these characters are very variable and identified material from collections in Europe often does not agree with the characters given by Engel (1930).

nigripennis. The spermathecae form long, thin, wrinkled tubes which are sclerotized in the distal part and covered with canaliculi. They taper to a long, thin end and form a large, loose, irregular coil. Valve strongly sclerotized, with long internal spines. Ejection apparatus short, relatively wide, with long processes to above the valve. There is a broad, ring-shaped sclerotization with pointed tubercles at the base. The ducts then continue in

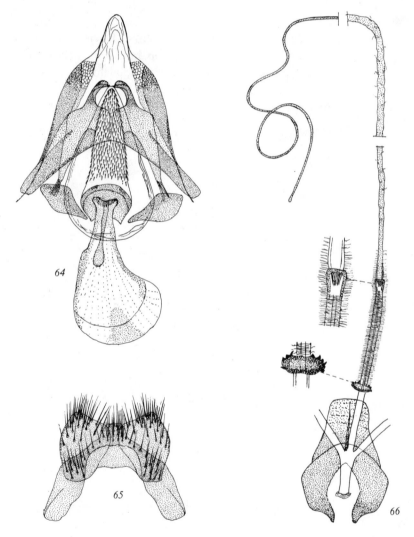

Figs. 64–66: *Holopogon nigripennis*
64. aedeagus; 65. proctiger; 66. spermatheca

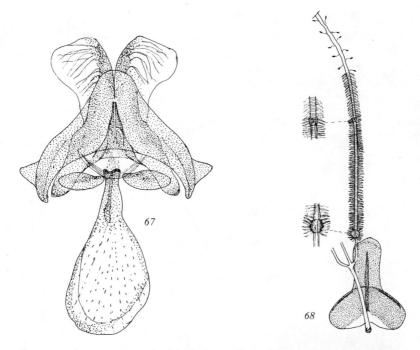

Figs. 67–68: *Holopogon dimidiatus.* 67. aedeagus; 68. spermatheca

smooth, narrow tubes into a short common duct. Furca plate-shaped, narrower apically, with a deep, rounded posterior indentation and a median longitudinal ridge.

Aedeagus broadly conical, with very wide apical opening. Endoaedeagus very large and broad, short, with distinct spines. There is a large ventral plate with membranous apex which is continuous with the hypandrium. Apodeme moderately wide, with a very large, rounded head.

fumipennis.   The spermathecae are very similar to those of *nigripennis*, except that the furca is of slightly different form.

Aedeagus resembles that of *nigripennis*, with minor differences. Endoaedeagus smaller and its connection with the apodeme narrow.

dimidiatus.   Spermathecae much shorter than in *nigripennis* and not forming a coil. Canaliculi distinct, with a large reservoir. Valve much smaller and ejection apparatus much longer and narrower than in *nigripennis*. Processes of ejection apparatus very long. Basal sclerotization small and rounded, the ducts proximal to it very narrow, and the common duct also longer and narrower. Furca broad at the base, apical part much narrower.

Aedeagus very short, broad at the base, narrowly conical apically, situated far posteriorly on the ventral plate, which is divided into two large, wing-shaped processes that extend far beyond the end of the aedeagus. Endoaedeagus small. Head of apodeme narrow.

Holopogon sp.   Spermathecae in general as in *dimidiatus*, but the ejection apparatus is much shorter and the basal sclerotization is rounded-conical, with distinct, large tubercles. Furca with broad, rounded basal part and narrow apical part.

Aedeagus broadly conical, without ventral plate. Endoaedeagus short, with spines.

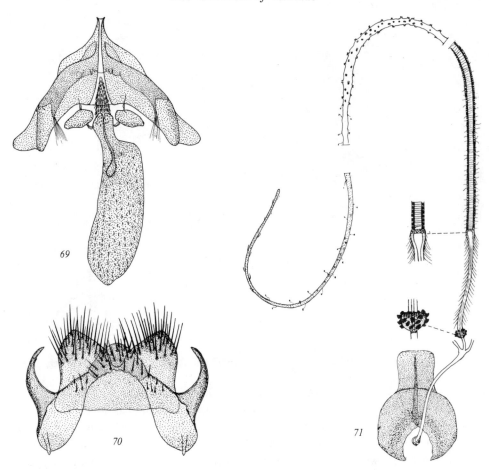

Figs. 69–71: *Holopogon* sp. 69. aedeagus; 70. proctiger; 71. spermatheca

The gonopods and proctiger of *Holopogon* show distinct differences in some of the species. Those of *nigripennis* and *fumipennis* are very similar. The proctiger is broad, without differentiations in three of the species examined, but with long, curved, horn-shaped processes in *Holopogon* sp. There are also distinct differences in the form of the epandrium and hypandrium.

H. nigripennis and *fumipennis* are apparently closely related and cannot be distinguished with certainty by the genitalia. However, some specimens may not be correctly identified, and more material should be examined.

Amphisbetetus (?) dorsatus, favillaceus and an undescribed species (Figs. 72–73)

The spermathecae of *Amphisbetetus* sp. are very long, thin tubes which are sclerotized in the distal part, forming a dense, regular spiral with 4–5 turns and clubbed end. Ducts markedly wider and thick-walled before the valve, which is conical and weakly sclerotized. Ejection apparatus short, with short processes. Furca plate-shaped, triangular, with a

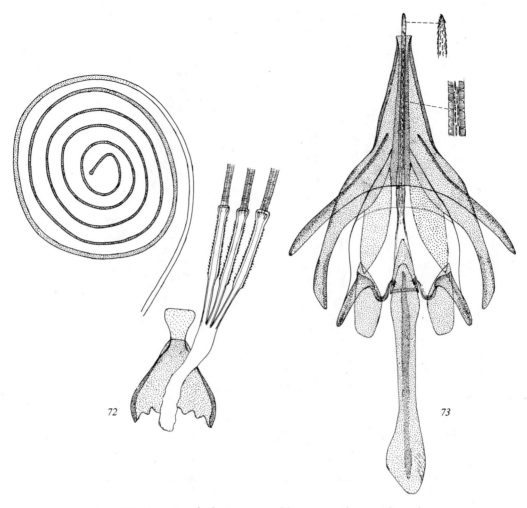

Figs. 72–73: *Amphisbetetus* sp. 72. spermatheca; 73. aedeagus

narrow, short, triangular apodeme, its wide side apical. Spirals of *favillaceus* thicker, valve larger and more strongly sclerotized.

Aedeagus of *Amphisbetetus* sp. conical, with well marked endoaedeagus, with pores and spines. Head of apodeme conical, pointed, connected with the endoaedeagus. Aedeagus of *dorsatus* similar, that of *favillaceus* more broadly conical.

Hypandrium of different form in the three species. Apical process of gonopods of *Amphisbetetus* sp. long, with curved, pointed end. Dististylus short, triangular. Apical process of gonopods of *favillaceus* similar, but thicker, dististylus smaller and narrower.

Sisyrnodytes major, nilicola and an undescribed species (Figs. 74–78)

nilicola. The spermathecae form a small spiral with 2–3 turns. There are distinct tubercles at the base of the canaliculi. The ducts become wider towards the base. Valve conical, distinct. Furca V-shaped, with a short, rectangular apodeme.

Aedeagus slender, conical. Endoaedeagus distinct, with spines and platelets. Apodeme narrow, with pointed head.

47

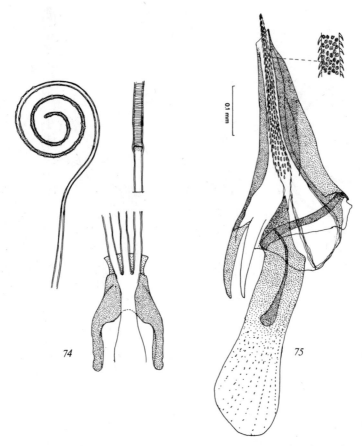

Figs. 74–75: *Sisyrnodytes nilicola.* 74. spermatheca; 75. aedeagus

Apical process of gonopods oblong, with short, pointed, curved end. Dististylus short, broad, rounded.

Sisyrnodytes sp.   Spermathecae resembling those of *nilicola*, but spiral with more turns, ejection apparatus and common duct longer. Furca wider, with a short, narrow, nearly rectangular apodeme.

Aedeagus very long and slender, about three times as long as that of *nilicola* (the species is only slightly larger than *nilicola*), uniformly wide in the basal two-thirds, tapering in the apical third. Apodeme short, its head with a short, pointed process. Endoaedeagus distinct, with long, lanceolate spines with a thin base.

Apical process of gonopods broad, truncate, with irregularly jagged margin. Dististylus short, oblong, with rounded end.

major.   The spermathecae form a thick, slightly curved, sclerotized tube which tapers to a thin, curled end. Ducts sclerotized, moderately long, thicker and striated before the valve, which is weakly sclerotized. Ejection apparatus slender. Common duct short and wide. Furca U-shaped, with wide lateral and apical parts. Apodeme narrow, rectangular.

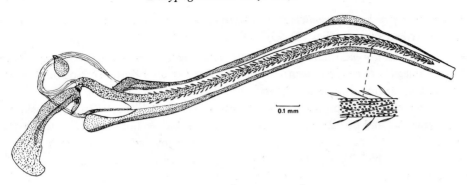

Fig. 76: *Sisyrnodytes* sp., aedeagus

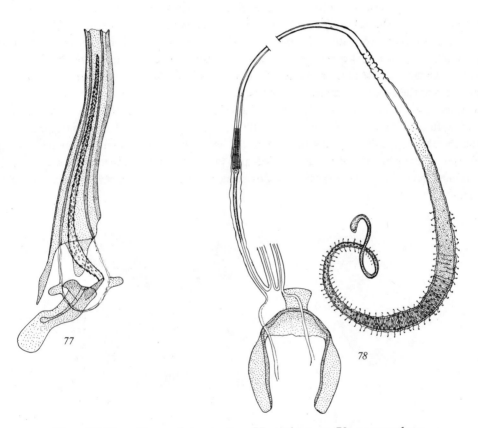

Figs. 77–78: *Sisyrnodytes major*. 77. aedeagus; 78. spermatheca

Aedeagus tubular, nearly straight, only slightly tapering apically, with wide opening. Endoaedeagus distinct, with platelets and spines. Apodeme short and narrow, its head wide, with a pointed process.

Apical process of gonopods thick, deeply bifid. Dististylus long, slender, with hooked end.

Habropogon aegyptius, appendiculatus, longiventris, malkovskii, spissipes, striatus and three undescribed species (Figs. 79–93)

The males of the *appendiculatus* group, with characteristic appendages on the pretarsus of their midlegs, are easily recognized, but the females, and both sexes of other species, are difficult to distinguish according to external characters.

The spermathecae of all species examined form ovoid or club-shaped capsules, which are darkly pigmented and of different size in the various species (all spermathecae are drawn to the same scale). The ducts are very narrow and are covered more or less densely with long canaliculi, which have a terminal reservoir of distinct form in some species. They are shorter and less numerous on the capsules. The posterior part of the ducts before the valve is without canaliculi. They are more densely arranged and longer on the ducts in *striatus*. All species examined, except species no. 1, have a strongly sclerotized valve which is situated far posteriorly. The valve is of different form in every species examined. The posterior part of the ducts of *longiventris* and of species no. 1 has internal spines and forms wide sacs with striated walls in *longiventris* and in species no. 2. The valves of some species (*appendiculatus, striatus*) continue in a sclerotized tube to the opening into the vagina. The furca is U-shaped, with an apodeme of different form in the various species. It has broad processes at the inner margin in species no. 1.

Aedeagus simple, conical, slightly curved, but of different form in the various species. It has a wide, funnel-shaped end in *longiventris*. It is regularly conical and slender in species no. 1. The sheath is wide distally and the end of the aedeagus is short and narrow in *aegyptius*. The pump chamber is wide and rounded and the head of the apodeme is wide and has a pointed process in some species. Endoaedeagus slender, weakly sclerotized, with a few spines.

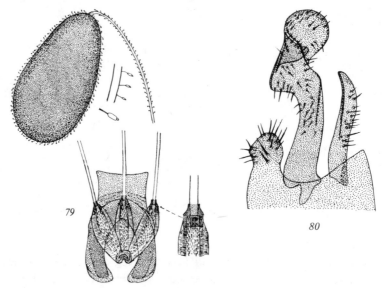

Figs. 79–80: *Habropogon longiventris*
79. spermatheca; 80. apical processes of gonopods and dististylus

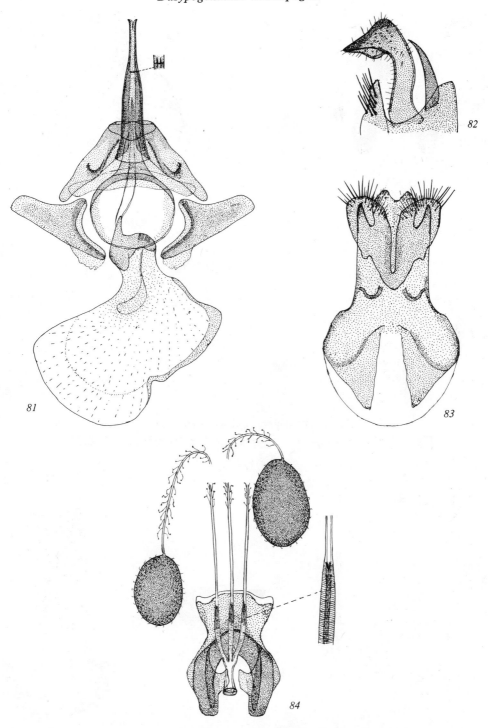

Figs. 81–84: *Habropogon* species no. 1. 81. aedeagus; 82. apical processes of gonopods and dististylus; 83. proctiger; 84. spermatheca

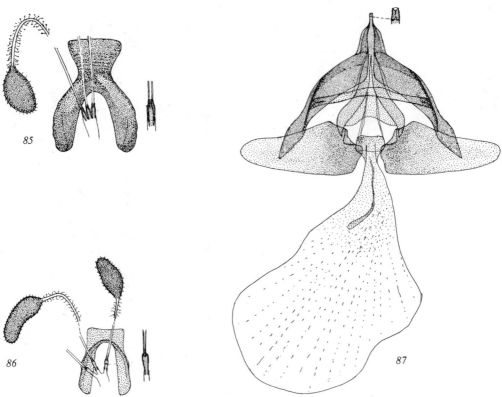

Fig. 85: *Habropogon appendiculatus*, spermatheca
Figs. 86–87: *Habropogon aegyptius*. 86. spermatheca; 87. aedeagus

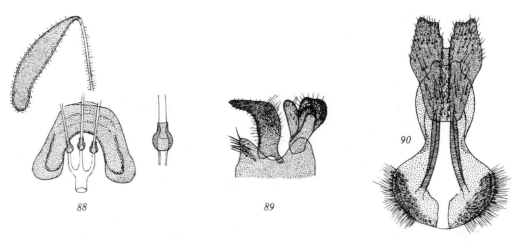

Figs. 88–90: *Habropogon malkovskii*. 88. spermatheca; 89. apical processes
of gonopods and dististylus; 90. proctiger

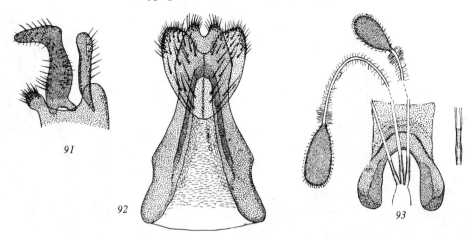

Figs. 91–93: *Habropogon striatus*
91. apical processes of gonopods and dististylus; 92. proctiger; 93. spermatheca

The apical process of the gonopods and the dististylus are of distinct form in most species. The end of the dististylus is curved dorsally and its form can be recognized only in separately mounted genitalia. The proctiger bears apical lateral processes in *longiventris* and species no. 1 and has a specific form in some other species. The apical process of the gonopods is blade-shaped or narrow in most species, but has a complicated form, with bulges and spines, in *malkovskii*.

Anarolius sp. (Figs. 94–95)
The spermathecae form long, sclerotized tubes which are thicker in the distal part and taper into a long, thin end which forms a loose coil. Openings of canaliculi large. Ducts wide, striated near the valve, which is conical and weakly sclerotized. Ejection apparatus with fine processes and a basal, rounded sclerotization with tubercles. Common duct short. Furca rectangular, with wide arms and a broad apodeme.
Aedeagus conical, simple, long, with wide opening. Sheath with two pointed denticles near the end. Endoaedeagus distinct, with spines.
Apical process of gonopods short, wide, plate-shaped, with a pointed inner process. Dististylus broad, pointed, with curved ventral margin and a ridge on the surface.

Pycnopogon mixtus, fasciculatus (Fig. 96)
The reservoir of the spermathecae forms a wide sac of distinct form which tapers into a long, thin, recurved process. Openings of canaliculi distinct. Ducts long, sclerotized, narrow, of similar structure as the reservoir. Posterior part of ducts delicate, striated, and with fine processes. Valve conical, weakly sclerotized. Ejection apparatus relatively short, with fine processes, ending in a sclerotized ring with tubercles, as in *Anarolius*. Common duct short. Furca rectangular and U-shaped, with broad apical apodeme.
Aedeagus simple, conical, with slightly funnel-shaped end. Endoaedeagus distinct, with spines.
Gonopods with a narrow, truncate ventral process and a broad, black, truncate dorsal process. Dististylus resembling that of *Anarolius*, but narrower. The male genitalia of

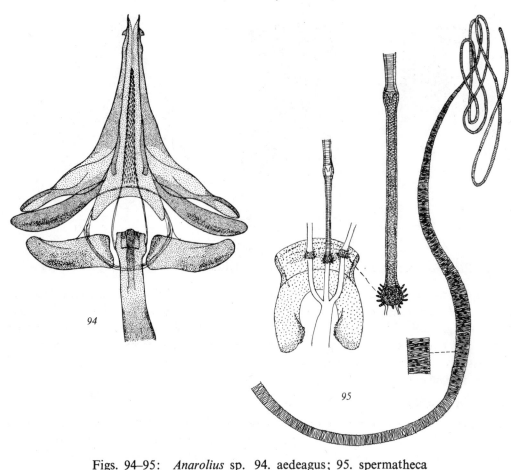

Figs. 94–95: *Anarolius* sp. 94. aedeagus; 95. spermatheca

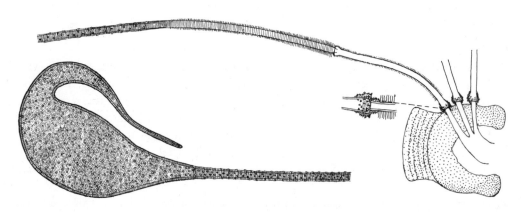

Fig. 96: *Pycnopogon mixtus*, spermatheca

fasciculatus are similar, but with specific differences. The aedeagus is shorter and its apex wider. The dististylus is wide at the base and the apical part is narrow.

Crobilocerus megilliformis and an undescribed species

The spermathecae form long, sclerotized tubes which are wider in the distal part and taper to a long, thin end. They become thicker before the valve, which is broadly conical. Ejection apparatus slender, without processes, with a large basal sclerotization with large, pointed tubercles, as in *Pycnopogon*, but larger. Furca U-shaped, with broad arms.

Aedeagus broadly conical, with wide, funnel-shaped end, resembling that of *Heteropogon* sp. near *ornatipes*. Endoaedeagus distinct, with spines.

Gonopods and dististylus resembling those of *Pycnopogon*. Hypandrium of *megilliformis* broadly triangular, with uniform setae in the apical part, that of the new species distinctly trapezoidal, with two tufts of setae at the sides of the apical margin.

Heteropogon nubilus, ornatipes, species no. 1 (near *ornatipes*), *pyrinus* (Figs. 97–101)

The spermathecae of species no. 1 form very long, sclerotized tubes which are wider and striated in the distal part and taper to a thin end. They are of granulate structure further on. Openings of canaliculi distinct. Ducts of similar structure in their greater part, delicate near the valve, which is conical and weakly sclerotized. Ejection apparatus long, with fine processes and a basal sclerotization with tubercles, as in *Anarolius* and *Pycnopogon*. It continues in short, wide ducts. Furca U-shaped, with broad apodeme.

Aedeagus of *ornatipes* short, broad, ending in three apical flanges. Endoaedeagus broad, conical, with spines. Apodeme with broad, rectangular head and large sensillae. Aedeagus of species no. 1 similar, but narrower apically. Aedeagus and spermathecae of *pyrinus* similar, but furca of different form, short, with very wide arms.

Gonopods of *ornatipes* with a narrow ventral and broadly truncate dorsal process. Dististylus broad in basal half, narrow in apical half.

nubilus.   The spermathecae are different. They form a sclerotized spiral with blunt end. Valve sclerotized, conical. Ejection apparatus with fine processes, continuing in wide, thin-walled ducts which end separately. Basal sclerotization absent. Furca narrowly triangular, with a median slit.

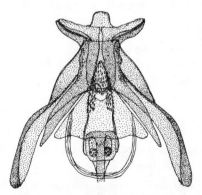

Fig. 97:   *Heteropogon ornatipes* (Europe), aedeagus

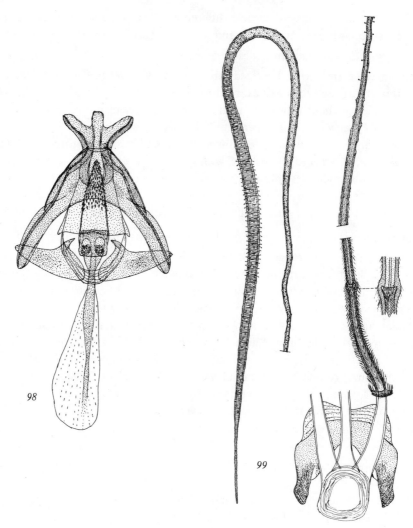

Figs. 98–99: *Heteropogon sp.* (near *ornatipes*)
98. aedeagus; 99. spermatheca

Aedeagus slender, its apical part long, tubular, curved. Apodeme with long, pointed head. Endoaedeagus with platelets and spines.

Gonopods with short, broadly rounded dorsal and narrow, cylindrical, truncate ventral process. Dististylus narrow, curved. Hypandrium very long, reaching far beyond the end of the gonopods, bifid at the end.

The genitalia of *nubilus* differ so markedly from those of *ornatipes* and *pyrinus* that it is doubtful whether *nubilus* should be retained in the genus *Heteropogon*. *nubilus* also differs markedly in external characters from species of the *manicatus* group. The genus *Heteropogon* seems artificial and in need of revision.

The genera *Heteropogon*, *Anarolius*, *Pycnopogon* and *Crobilocerus* form a well defined group according to the structure of the spermathecae.

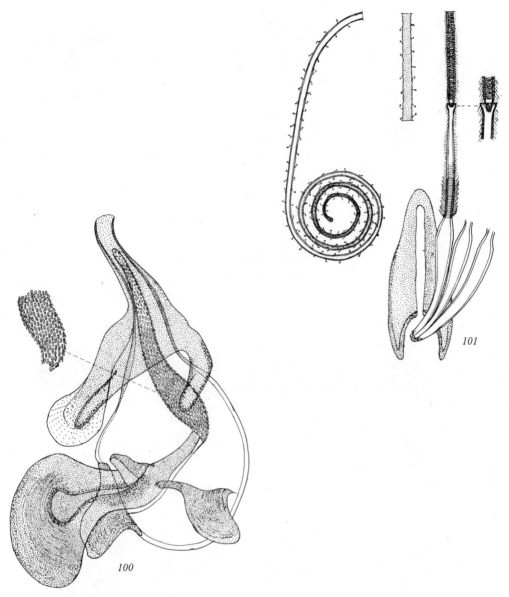

Figs. 100–101: *Heteropogon nubilus.* 100. aedeagus; 101. spermatheca

Anisopogon hermanni, parvum (Figs. 102–108)

hermanni. Spermathecae with very long, sclerotized tubes which taper to a fine end and do not form a spiral. Openings of canaliculi distinct. Ducts sclerotized to the valve, which is weakly sclerotized. Ejection apparatus long, delicate, ending in a sclerotized cylindrical part which continues in wide, crinkled, membranous tubes. Furca broad, rhomboidal, its posterior ends nearly contiguous.

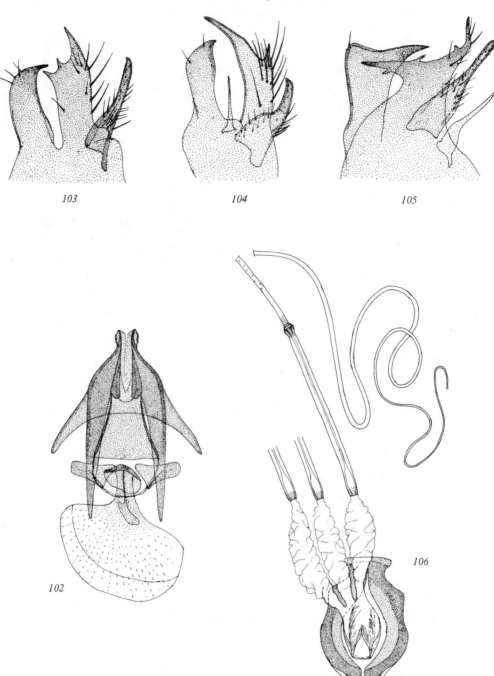

Figs. 102–106:   *Anisopogon hermanni*
102. aedeagus; 103. apical processes of gonopods and dististylus;
104. same, variation; 105. same, variation or different species (?);
106. spermatheca

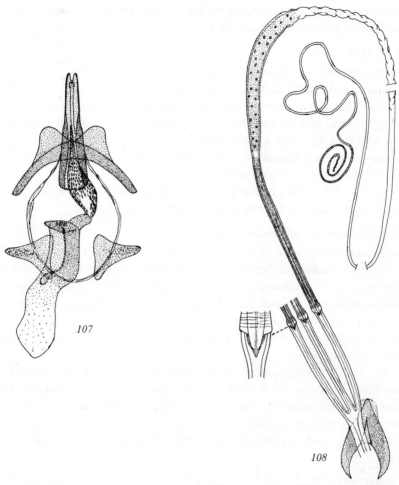

Figs. 107–108: *Anisopogon parvum*. 107. aedeagus; 108. spermatheca

Aedeagus short, conical, broad, with fine lateral serrations at the end. Apodeme with wide head. Endoaedeagus not recognizable.

The apical process of the gonopods shows marked variations in form (Figs. 103–104). Its lateral processes are either very long, diverging or not, or short. The number of small denticles also varies, and denticles are absent in some specimens. Some specimens which are indistinguishable from *hermanni* in external characters show such marked differences in the form of the dististylus and apical process that they may belong to a different species (Fig. 105).

parvum. The spermathecae form a small spiral, the ducts are crinkled further on, continue in a wide sclerotized part and then become narrow again and striated before the conical valve. Furca narrow, V-shaped.

Aedeagus with long, narrow end. Sheath with two rounded apical processes. Endoaedeagus distinct, with oblong platelets. Basal apodemes triangular.

Gonopods with an outer, pointed apical process and a broad, inner, plate-shaped process with rounded end and a high, vertical median ridge. Dististylus broad, S-curved, with rounded end and a lateral corner near the base.

Scylaticus sp. (Fig. 109)

The spermathecae form very wide, sclerotized tubes in a loose spiral with 1–2 turns and broadly rounded end. Basal half of spiral with short longitudinal folds. Openings of canaliculi large. The reservoir narrows abruptly into very thin, long ducts. Valve strongly sclerotized, cylindrical, with an internal cone. Ejection apparatus with fine processes. Furca rectangular, with wide arms and a broad apodeme.

Aedeagus simple, conical. Sheath with wide apical and ventral bulges. Apodeme with a very wide head. Endoaedeagus with spines.

Gonopods with a curved, thin lateral process, a thick, club-shaped apical process and a short, curved dorsal process with long setae. Dististylus thick, curved.

Ancylorrhynchus glaucius, rufocinctus (Figs. 110–111)

Spermathecae resembling those of *Scylaticus* in general, forming a loose spiral with 2–3 turns and very wide tubes, with longitudinal folds in the basal part in *glaucius*. The spiral narrows abruptly into a narrow duct. The ducts are very long and arranged in a spiral with several turns around its wide part. Valve cylindrical in *glaucius*, narrower and ring-shaped in *rufocinctus*. The tubes of the spiral are much narrower in *rufocinctus*. Furca of *glaucius* nearly rectangular, with lateral processes in the middle and posteriorly, with lateral corners in *rufocinctus*.

Aedeagus of *glaucius* simple, conical, with short apical part. Sheath with two rounded apical processes. Apodeme with wide, rectangular head. Endoaedeagus with spines. Aedeagus of *rufocinctus* narrower, sheath without apical processes.

Apical process of gonopods of *glaucius* with a lateral rounded process. Dististylus long, thick, with truncate end. Apical process and dististylus of *rufocinctus* shorter, dististylus curved.

Jothopogon leucomallus (Fig. 112)

The spermathecae form a thick spiral with 4–5 turns, tapering in the distal part, transversely striated. Openings of canaliculi distinct. Ducts wide, ending in a broad, cylindrical valve and continuing in smooth, thick-walled tubes. Furca U-shaped, with widened posterior ends with reticulate structure.

Aedeagus apparently 3-pronged, but in fact consisting of a tube with dorsal serrations at the base and two ventral processes parallel to the tube and nearly as long as it. They have dorsal serrations near the apex. Base of sheath with a large, nearly semicircular, flattened process above the base of the aedeagus. Apodeme slender, head T-shaped. Endoaedeagus short, with spines.

Apical process of gonopods obliquely truncate, situated on the inner side of the gonopod. Dististylus long, with club-shaped, widened end.

Cyrtopogon lateralis, ruficornis (Figs. 113–115)

The spermathecae of *ruficornis* form a large, thick, sclerotized spiral with 2–3 turns. Openings of canaliculi distinct. The spirals narrow abruptly into a narrow duct with dense, long canaliculi. Posterior part of ducts without canaliculi. Valve weakly marked. Furca truncate-

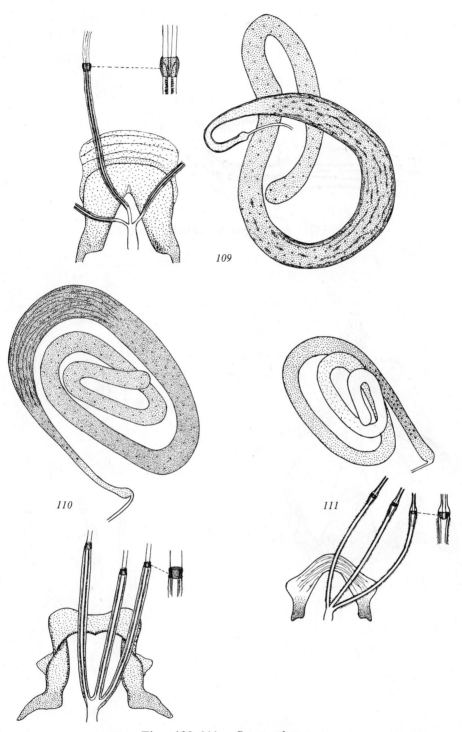

Figs. 109–111: Spermatheca
109. *Scylaticus* sp.; 110. *Ancylorrhynchus glaucius*; 111. *A. rufocinctus*

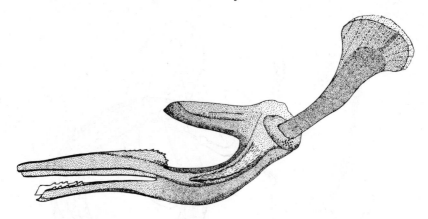

Fig. 112: *Jothopogon leucomallus*, aedeagus

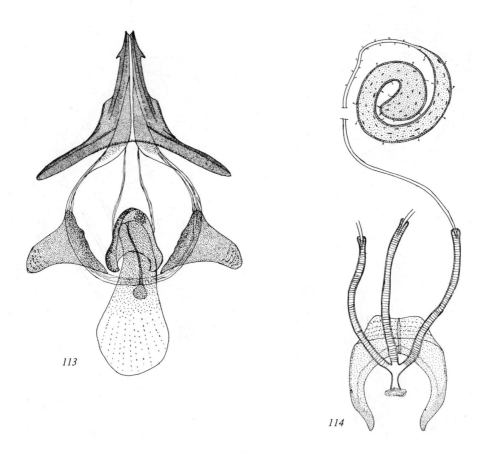

Figs. 113–114: *Cyrtopogon lateralis*. 113. aedeagus; 114. spermatheca

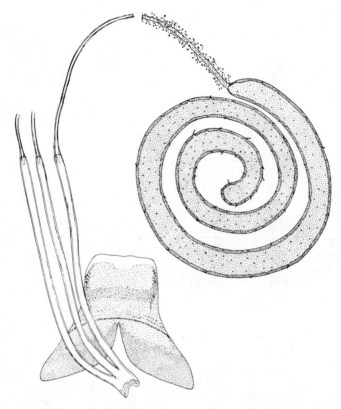

Fig. 115: *Cyrtopogon ruficornis*, spermatheca

V-shaped, with a small posterior incision. Spiral of *lateralis* smaller, with only 1–2 turns. Ducts with few canaliculi. Valve conical. Furca nearly ring-shaped, arms tapering posteriorly, apical part wide.

Aedeagus of *lateralis* conical, sheath with two ventral denticles in the apical part. Pump chamber rounded, very wide. Apodeme with very large head.

Gonopods with a broad, triangular process, its wide side apical, and a narrow, curved process at the apex. Dististylus straight, narrow, tapering.

Mimoscolia oberthurii (Figs. 116–117)

Spermathecae with wide tubes which form a loose spiral with 1–2 turns. They taper gradually into narrow, sclerotized ducts. Posterior part of spiral and adjacent part of ducts with openings of canaliculi on tubercles. Ducts narrow to the valve, which is large and conical, with internal spines. Ejection apparatus thick, with internal spines, continuing in narrower ducts into a short common duct. Furca broadly U-shaped, arms tapering posteriorly.

Aedeagus broadly conical, with two lateral, pointed corners near the end. Sheath with a large ventral bulge. Pump chamber very large. Endoaedeagus with pores and granules. Apodeme with large, rounded head.

Gonopods with two apical processes, one broad, triangular, the other narrow, tapering. A plate-shaped process near the base of the gonopod. Dististylus club-shaped.

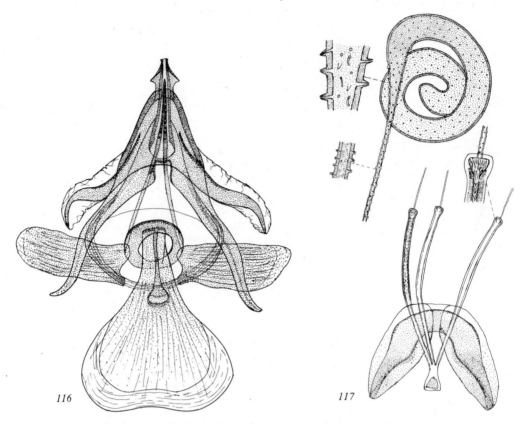

Figs. 116–117: *Mimoscolia oberthurii*. 116. aedeagus; 117. spermatheca

Dioctria atricapilla, oelandica, rufipes, valida (Figs. 118–129)

Spermathecae with long, thin tubes which form a sclerotized, flattened spiral ending in a small club. Ducts very long, crinkled in the posterior part, widening markedly before the valve, which is weakly sclerotized. Ejection apparatus narrow, long, with fine processes. Common duct short. Furca consisting of two lateral bars which form an angle, connected by a membrane. Only minor variations in the different species.

valida.   Aedeagus conical, its end tubular. Sheath with two lateral denticles in the apical part. Endoaedeagus with spines.
Epandrium with two slender, pointed apical processes. Gonopods with a straight, narrow apical process and a shorter, curved, pointed process with three setae. Dististylus broad at the base, tapering in the apical half, with a ventral apical point.

atricapilla.   Aedeagus with two lateral denticles near the apex. Sheath abruptly widening into lateral corners. Endoaedeagus granulate.
Epandrium with two conical lateral processes. Dististylus broad with pointed ventral end.

rufipes.   Aedeagus conical, with two large ventral, truncate processes. Endoaedeagus with granules and spines.
Epandrium with rounded, lateral apical processes. Dististylus slender in apical half, with rounded end. Apical process of gonopods curved, pointed.

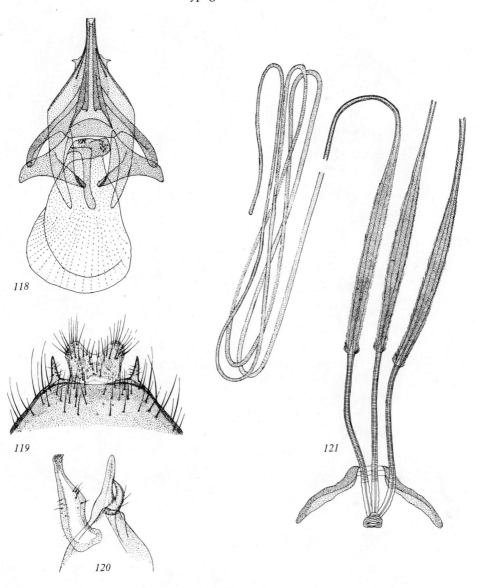

Figs. 118–121: *Dioctria valida*
118. aedeagus; 119. end of epandrium and proctiger;
120. apical processes of gonopods and dististylus; 121. spermatheca

oelandica. Aedeagus resembling that of *rufipes*, but ventral processes small, triangular. Endoaedeagus granulate and with long, thin spines in the basal part.

Epandrium with thick apical processes. Dististylus slender in apical half, with wide, flattened, rounded end.

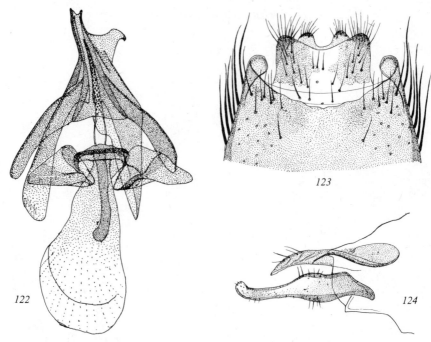

Figs. 122–124: *Dioctria rufipes*. 122. aedeagus, oblique;
123. end of epandrium and proctiger;
124. apical process of gonopod and dististylus

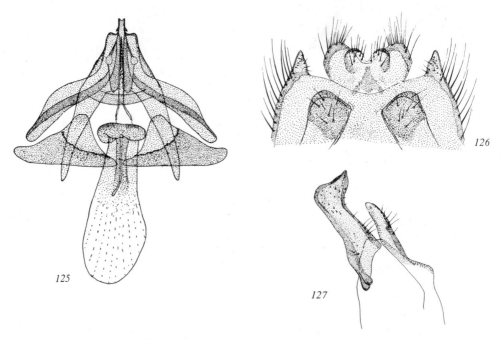

Figs. 125–127: *Dioctria atricapilla*
125. aedeagus; 126. end of epandrium and proctiger;
127. apical process of gonopod and dististylus

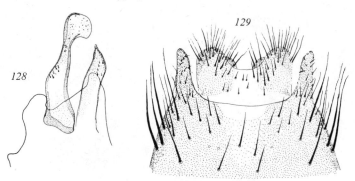

Figs. 128–129: *Dioctria oelandica*
128. apical process of gonopod and dististylus;
129. end of epandrium and proctiger

Stichopogonini

Stichopogon albellus, albofasciatus, chrysostoma, elegantulus, flaviventris, inaequalis, scaliger, schineri (Figs. 130–139)

The spermathecae form a small, sclerotized spiral with several turns and distinct tubercles at the openings of the canaliculi. The ducts widen towards the valve in some species (*scaliger, albellus*). They are nearly club-shaped in *albellus*. Furca more or less V-shaped, of varying

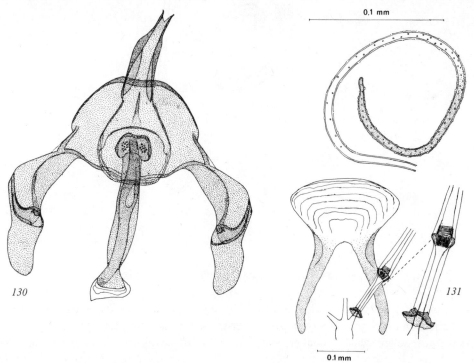

Figs. 130–131: *Stichopogon schineri*. 130. aedeagus; 131. spermatheca

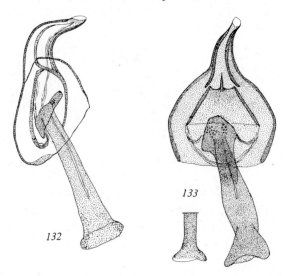

Figs. 132–133: *Stichopogon chrysostoma*
132. aedeagus, lateral; 133. same, ventral

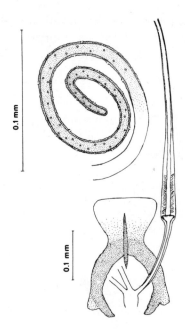

Fig. 134: *Stichopogon albellus*, spermatheca

form in the different species. The spiral of *schineri* is similar, but the valve has internal spines. Ejection apparatus with a wide, flower-shaped, basal sclerotization in *schineri* and *albofasciatus*. Sclerotization smaller, more rounded in *flaviventris*, absent in the other species.

Aedeagus of most species simple, conical, with curved or S-curved end. Apodeme with

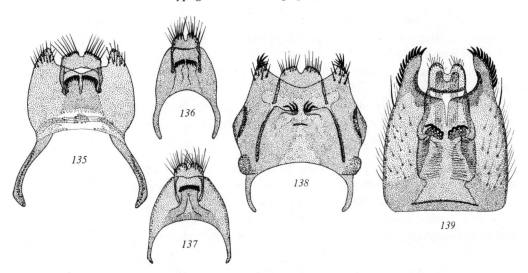

Figs. 135–139: Epandrium and proctiger of *Stichopogon*. 135. *S. scaliger*;
136. *S. elegantulus*; 137. *S. albellus*; 138. *S. schineri*; 139. *S. albofasciatus*

large, rounded head and a transverse bar at the free end, so that it appears T-shaped in
dorsal view. Aedeagus of *schineri* slender, sheath abruptly widening posteriorly.

Dististylus of all species examined short, thick, rounded, but with specific differences,
particularly large in *albofasciatus* and *schineri*.

The proctiger of all species examined bears a characteristic armature of spines in one row
which may be divided in the middle. The number and form of the spines vary in the different
species. They are geniculate in *schineri* and *albofasciatus*. The epandrium has two apical
processes with spines at the apex in *schineri* and two lateral rows in *albofasciatus*.

Rhadinus sp. near *laurae, mesasiaticus* (Fig. 140)

Spermathecae with long, thin tubes which form a flattened spiral or coil, sclerotized only
in a small distal part, tapering to a long, thin end. Ducts thick-walled posteriorly. Valve

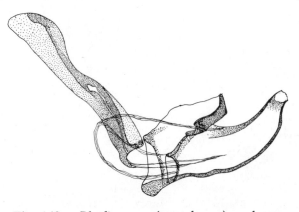

Fig. 140: *Rhadinus* sp. (near *laurae*), aedeagus

narrow, conical, weakly sclerotized. Ejection apparatus very short. Furca V-shaped, with a wide, triangular apodeme, the broad side apical.

Aedeagus simple, conical, nearly tubular in the basal part, with narrow, upturned end. Apodeme slender, head with a small, pointed process. Endoaedeagus indistinct, without spines.

Apical process of gonopods of complicated form, with two black, pointed tubercles. Dististylus short, thick, with pointed end, with specific differences in the two species.

Lasiopogon cinctus (Figs. 141–143)

The spermathecae form long, tapering, thin-walled tubes that expand proximally into a wide bulb and then narrow again before the valve, which is weakly sclerotized and has internal spines. Furca long, V-shaped, with a median ridge.

Aedeagus slender, conical. Sheath with two long, blade-shaped processes parallel to the end of the aedeagus, so that it appears 3-pronged. Pump chamber very large, sclerotized

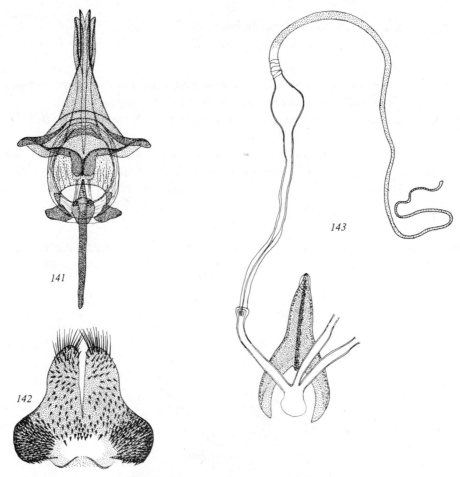

Figs. 141–143: *Lasiopogon cinctus*
141. aedeagus; 142. proctiger; 143. spermatheca

except around the head of the apodeme, which is rounded and has a long, pointed process. Endoaedeagus distinct, slender, granulate. Basal plates small, black, oblong.

Gonopods short, truncate, with a dense brush of setae at the ventral margin. Dististylus black, broad at the base, with a triangular lateral process and a narrower, truncate end which is wider at the apex. Posterior apodeme of gonopods very large, fan-shaped. Hypandrium rounded, fused with the gonopods in its larger basal part.

Proctiger with two wide, rounded lateral bulges in the proximal part which are covered with spines and setae. The spines are particularly numerous and strong on the lateral bulges.

Damalini Hull (Xenomyzini Oldroyd)

Holcocephala abdominalis, alboatra, affinis, calva, oculata, Holcocephala sp. (Figs. 144–159)

The spermathecae form a rounded or flattened spiral. Ducts very long, so that the spermathecae extend through the whole abdomen in some species. They are shorter, with a rounded spiral, in others. Reservoir of canaliculi very large in some species, rounded or oblong. Proximal part of ducts wide, with numerous pointed processes. Ejection apparatus of varying form, with a sclerotized valve with external spines and with reticulate structure, mainly in the apical part. Ducts ending in a sclerotized ring. Furca broadly U-shaped, of varying form, but always with a more or less wide apical part with pointed lateral processes.

Aedeagus fork-shaped, with three tubular, functional prongs. In the different species the form of the aedeagus varies markedly in length, the form and length of the prongs and the width of the sheath. Basal plates small, triangular in *abdominalis*, with wide, rounded

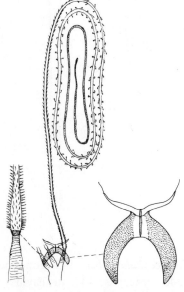

Fig. 144: *Holcocephala abdominalis*, spermatheca

71

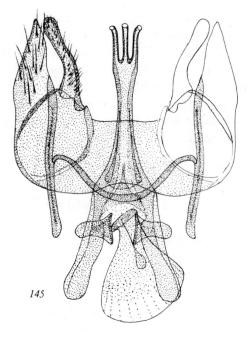

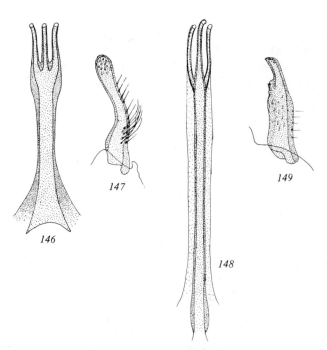

Figs. 145–147: *Holcocephala calva.* 145. gonopods and aedeagus;
146. aedeagus; 147. dististylus
Figs. 148–149: *Holcocephala* sp. 148. aedeagus; 149. dististylus

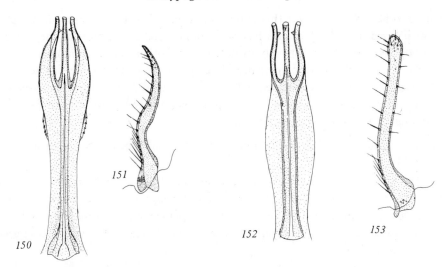

Figs. 150–151: *Holcocephala alboatra*
150. aedeagus; 151. dististylus

Figs. 152–153: *Holcocephala abdominalis*
152. aedeagus; 153. dististylus

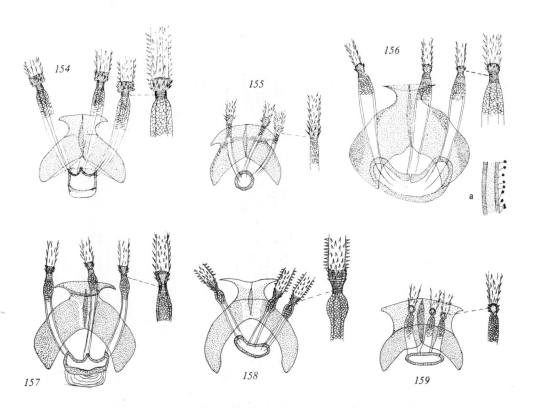

Figs. 154–159: Basal part of spermathecae and furca
154. *Holcocephala calva*; 155. *Holcocephala* sp.; 156. *H. affinis*; (a) part of
duct with canaliculi; 157. *H. alboatra*; 158. *H. abdominalis*; 159. *H. oculata*

lateral part in other species. So far this is the only genus of Dasypogoninae in which an aedeagus with three functional prongs has been found. The connection of the aedeagus with the gonopods also differs distinctly from that in all other Dasypogoninae examined. It is not effected by the posterior apodemes, but by a curved, transverse bar across the base of the aedeagus which is connected with the rod-shaped apodemes of the gonopods. The dististylus also differs markedly in form in the different species, as shown in the figures. Hypandrium partly fused with the gonopods.

Rhipidocephala caffra, obscurata, scutata and an unidentified species from Kenya (Figs. 160–169)

obscurata. The spermathecae form extremely long, thin tubes in a flattened spiral with four turns and a small spiral at the end. They extend through the whole abdomen and are several times longer than the whole insect. The ducts widen slightly before the valve and bear short processes. Ejection apparatus short, reticulate. Furca with a trapezoidal apical plate and thin lateral arms.

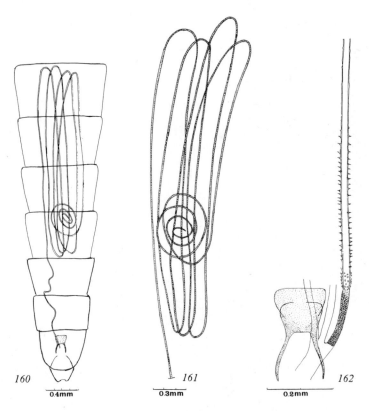

Figs. 160–162: *Rhipidocephala obscurata.* 160. spermatheca *in situ*; 161. spermatheca; 162. base of ducts and furca

Fig. 163:   *Rhipidocephala scutata,* furca and base of ducts

Fig. 164:   *Rhipidocephala caffra,* furca

Aedeagus slightly conical, curved at the base, with two apical, narrowly triangular processes. Basal plates funnel-shaped. Apodeme small, of irregular form. Hypandrium with two lateral processes and a long, black horn in the middle. Dististylus and gonopods of complicated form.

scutata.   Spermathecae similar, but furca narrower.

Aedeagus curved at the base, apical part wide, nearly tubular, with a wide opening and two long, dorsal lateral processes at the apex. Basal plates funnel-shaped. Hypandrium trapezoidal, with a membranous area before the apex and a tubercle instead of the horn in *obscurata.* Gonopods and dististylus resembling those of *obscurata,* with minor differences.

caffra.   Spermathecae similar, but without differentiations at the base, and ducts more irregularly coiled. Furca of similar form, but apical apodeme short and lateral arms much longer. Male not examined.

Species from Kenya.   Aedeagus shorter than in the preceding species, not curved, only slightly tapering, with denticles at the apex. Basal plates nearly rectangular. Apodeme large, with wide head. Gonopods and dististylus of characteristic form.

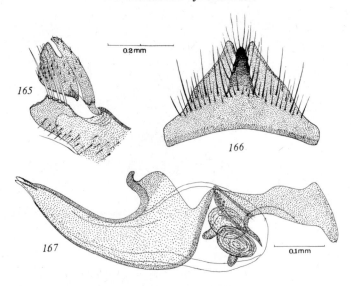

Figs. 165–167: *Rhipidocephala obscurata*
165. dististylus; 166. hypandrium; 167. aedeagus

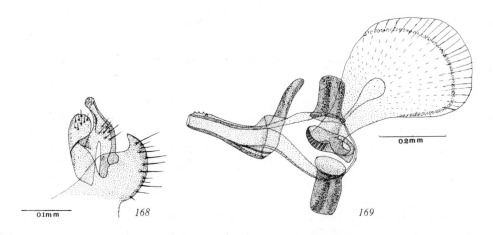

Figs. 168–169: *Rhipidocephala* sp. (Kenya). 168. dististylus; 169. aedeagus

Damalina sp. (Thailand) (Fig. 170)

The spermathecae form thick, sclerotized, recurved tubes with a small bulb at the end. The sclerotized part of the ducts passes without a recognizable valve into thin-walled ducts with long processes. These widen at the base into a bulb with a basal sclerotization with tubercles and continue in membranous ducts which widen markedly at the base and open behind the furca in wide openings with a thick, sclerotized frame. Furca U-shaped, with short, tapering lateral arms and wide apical part. Male not examined.

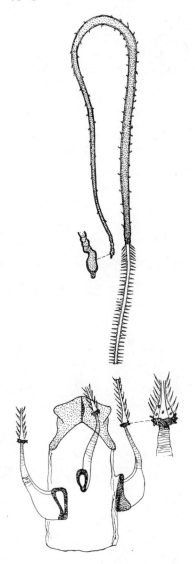

Fig. 170:   *Damalina* sp., spermatheca

Oligopogon hybotinus, species near *hybotinus* (Congo) and an unidentified species from South Africa (Figs. 171–174)

The spermathecae of the species from South Africa form a small spiral with 3–4 turns which is wider in the outer turn, and with distinct canaliculi. Ducts long, moderately thick, coiled. Proximal part of ducts striated, with processes. Valve weakly sclerotized, with internal spines and a basal ring-shaped sclerotization, continuing in short membranous ducts and a short common duct. Furca plate-shaped, with a posterior incision.

Species near *hybotinus* with similar spermathecae, but the tubes in the spiral are not widened in the outer turns.

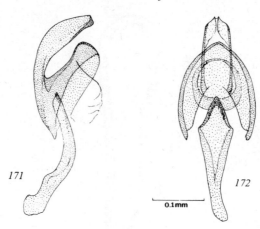

Figs. 171–172: *Oligopogon hybotinus*
171. aedeagus, lateral; 172. same, dorsal

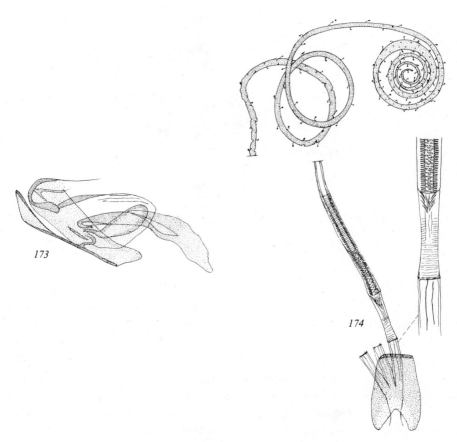

Figs. 173–174: *Oligopogon* sp. (South Africa)
173. aedeagus, lateral; 174. spermatheca

The aedeagus of *hybotinus* has a short, narrow, flattened apical part. The sheath forms a large bulge which extends to near the end of the apical part. Apodeme slender, curved, of irregular form, with wide, conical head and without a neck. Basal plates triangular. The aedeagus of the species from South Africa is similar, but the apical part is more distinctly flattened. Basal plates triangular, with rounded lateral end and a thin process connecting them with the pump chamber.

Damalis femoralis, taciturna (Figs. 175–181)

femoralis. The spermathecae form long, thick tubes in a flattened spiral. Ducts with relatively large, slit-shaped openings of the canaliculi, which have oblong reservoirs. Ducts membranous in proximal part. Valve weakly sclerotized. Ejection apparatus with short processes. Furca V-shaped, with narrow arms and rounded apex.

Aedeagus short, slightly conical, with a short, narrow apical part which forms an angle with the basal part. Basal plates rectangular, black, hollow, connected with the aedeagus by a narrow ligament. Endoaedeagus distinct, apodeme very large.

The gonopods are of very unusual form. There are two long, pointed apical processes, the inner process with setae at the apex. There is an inner, broadly truncate plate which is concave to the outside and bears two apical lateral processes. The plate is apparently fused with the gonopod. There is, however, a fascicle of tonofibrillae at its base, as on the dististylus of other genera. This is thus possibly a modified dististylus. There is a tuft of long, black setae on the inner corner of the gonopod which continues in an oblique row across it. Gonopods broadly fused at the base, but with a median suture. Hypandrium absent.

taciturna. The spermathecae form a small spiral with 2–3 turns and slightly club-shaped end. Openings of canaliculi on longitudinal tubercles in the thick part of the ducts. Canaliculi with oblong reservoirs. Valve distinctly sclerotized, conical. Ejection apparatus striated, with a proximal, cylindrical sclerotization, continuing in membranous tubes which widen markedly proximally. Furca V-shaped, with rounded apex and a rounded, plate-shaped keel in the middle of the lateral arms.

Aedeagus broadly conical, its apex markedly widened, with a posteriorly directed denticle at each side. Apical part with angularly bent apodemes on which a muscle is inserted. Apodeme with a dorsal, posterior, horizontal plate and with large, rectangular head. Basal plates fused with the sheath at the posterior lateral corners.

Gonopods completely fused, with a long dorsal apical process with 4–5 long, strong spines and with a triangular ventral process. Dististylus normal, with very broad basal part and with a narrow apical part with truncate end.

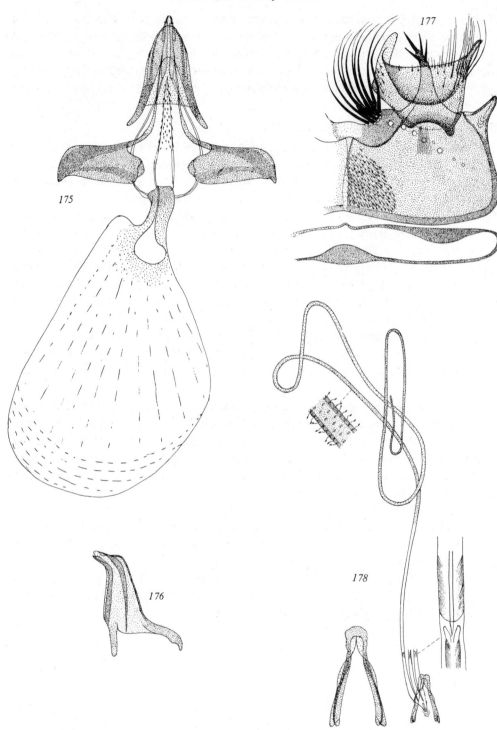

Figs. 175–178: *Damalis femoralis*. 175. aedeagus, dorsal; 176. same, lateral; 177. gonopod, reduced sternite and tergite 8; 178. spermatheca

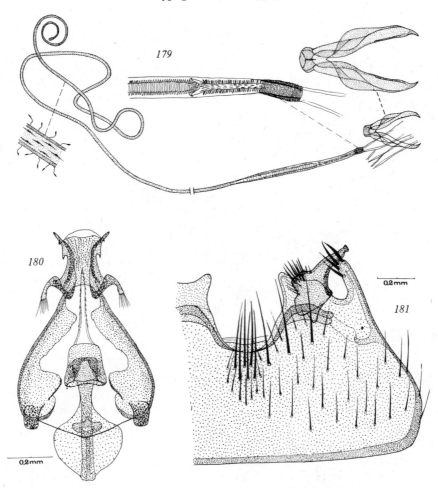

Figs. 179–181: *Damalis taciturna*
179. spermatheca; 180. aedeagus; 181. gonopod

Dasypogonini

Dasypogon diadema (Figs. 182–184)

The reservoir of the spermathecae forms wide, short, sclerotized, slightly curved tubes with rounded end. There are long spines inside the tubes at the openings of the canaliculi and fine external canaliculi. Ducts wide, also with canaliculi in the distal part. Valve not recognizable. Furca forming a nearly rectangular plate with a small posterior concavity.

Aedeagus conical, with a curved, tubular ending with transverse ridges, which is divided in the middle. Sheath with two long, ventral processes which project markedly beyond the end of the aedeagus and are curved inwards, so that their pointed ends meet. Apodeme broad, with large head.

Apical process of gonopods and dististylus blade-shaped, dististylus curved.

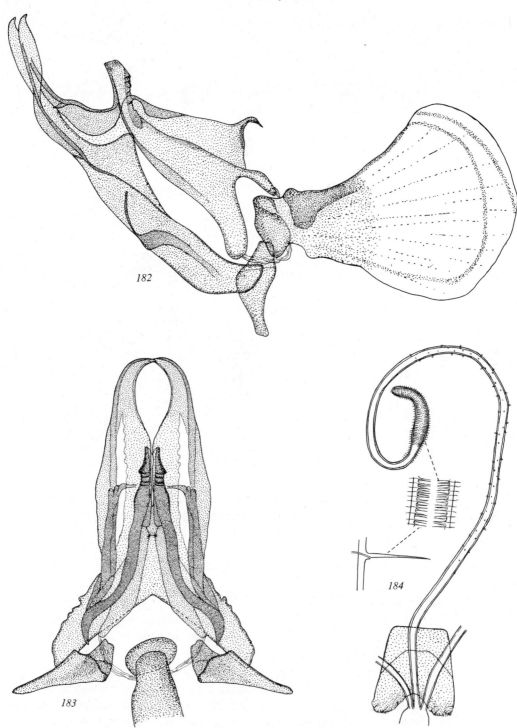

Figs. 182–184: *Dasypogon diadema*
182. aedeagus, lateral; 183. same, dorsal; 184. spermatheca

Saropogon aurifrons, distinctus, species no. 1 (near *elbaensis*), *jugulum, leucocephalus, longicornis, pittoproctus,* species no. 2 (near *pittoproctus*), *dasynotus, platynotus,* species no. 3 (near *platynotus*) and another four undescribed species (Figs. 185–215)

The spermathecae form large, sclerotized spirals with internal spines as in *D. diadema.* The number of turns of the spiral and the width of the tubes vary in the different species (1–2 turns in *aurifrons* and species nos. 4 and 6, 3–4 turns in species no. 1). The tubes of the reservoir are very wide in *aurifrons* and *distinctus,* and the ducts have a basal sclerotization. The ducts are wider and thick-walled in the posterior part and covered with musculature;

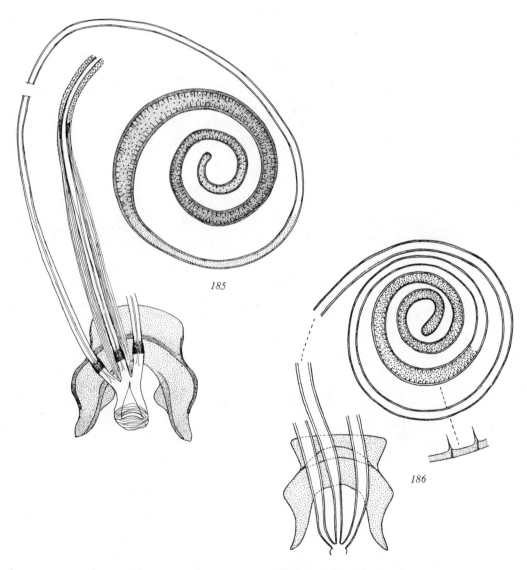

Figs. 185–186: Spermatheca of *Saropogon*
185. *S. distinctus;* 186. species no. 1 (near *elbaensis*)

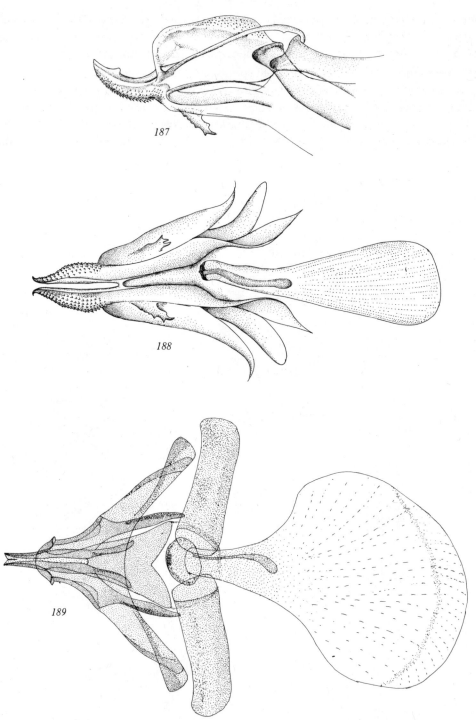

Figs. 187–189: Aedeagus of *Saropogon*. 187. *S. leucocephalus*, lateral;
188. same, ventral; 189. *S. aurifrons*

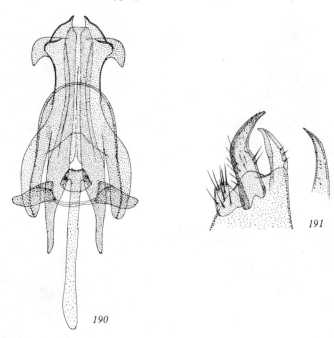

Figs. 190–191: *Saropogon jugulum*. 190. aedeagus;
191. apical processes of gonopod and dististylus

they apparently act as an ejection apparatus, but a valve is absent. Furca U-shaped, with narrow arms in some species, a wide or narrower apodeme in others, V-shaped in species no. 2, Y-shaped, i.e. angular, with a relatively long and narrow apodeme, in species no. 5. The aedeagus is conical, but shows marked differentiations, in all the species examined. The dististylus, and particularly the apical process of the gonopods, have a specific form in nearly every species examined.

distinctus.   Aedeagus conical, without denticles, but with two wide, angular processes near the apex on the dorsal side. These processes continue in lateral, curved ridges. Apical process of gonopods very broad, bifid, subapical branch rounded, apical branch narrow, curved, with rounded end. Dististylus slender, with pointed end. Proctiger rounded, without processes.

Species no. 1.   Aedeagus similar, but free end longer, tapering. Sheath with two semi-circular ridges near the end. Apical process of gonopods broad, bifid, resembling that of *distinctus.*

leucocephalus.   Aedeagus slender, slightly curved dorsally, tapering to a point, with two large denticles near the end. Sheath densely covered with denticles in the apical part. Ventral side of sheath with two large tubercles with irregular processes. Apodeme large, rounded. Proctiger without processes. Apical process of gonopods bifid, apical branch longer, curved, subapical branch short, the two branches about equally thick. Dististylus tapering, curved.

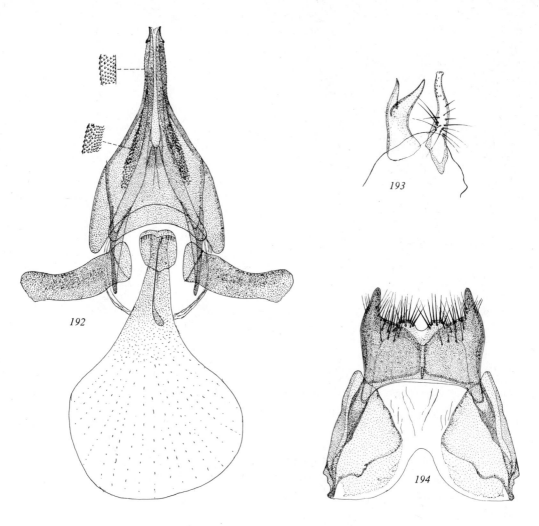

Figs. 192–194: *Saropogon* species no. 6
192. aedeagus; 193. apical process of gonopod and dististylus; 194. proctiger

aurifrons.   Aedeagus broadly conical, its apical part tubular. Sheath without denticles, with two apical semicircular ridges which extend to the apex. Proctiger wide, with two narrow, pointed apical processes at the lateral corners. Apical process of gonopods bifid, apical branch slender, short, curved, with rounded end, subapical branch broadly rounded.

jugulum (specimen from the type locality, Rhodes).   Aedeagus wide apically, without a free apical part. Sheath with two flattened, recurved, lateral processes at the apex. Apical processes of gonopods simple, blade-shaped, with pointed end. Dististylus slender. The aedeagus and the processes of the gonopods resemble those of *platynotus*. Proctiger with broadly triangular posterior lateral processes with rounded end and a lateral angle.

Species no. 6. This species closely resembles *jugulum* in external characters and was identified as *jugulum* by Engel (1930) and Efflatoun (1937). Comparison of the male genitalia with those of *jugulum* show that they differ markedly. Aedeagus slender, conical, with two

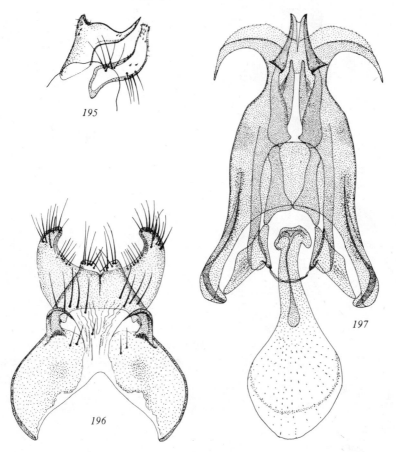

Figs. 195–196: *Saropogon longicornis*
195. apical process of gonopod and dististylus; 196. proctiger
Fig. 197: *Saropogon platynotus*, aedeagus

pointed lateral denticles near the apex. Ventral side of sheath densely covered with denticles except in a median stripe. Proctiger with long, pointed processes in the lateral posterior corners. Apical process of gonopods bifid, with long, pointed branches.

longicornis. Aedeagus resembling that of species no. 6, but its end is triangular in lateral view and the subapical denticles are much smaller. Proctiger with wide, rounded apical processes which are directed inwards and bear long setae. There are two rounded processes in the basal part and a row of setae distal to them. Apical processes of gonopods bifid, subapical branch short, apical branch long, pointed. Apical margin undulate. Dististylus strongly curved.

platynotus. Aedeagus with short, conical free end. Sheath with two large, flattened, apical dorsal processes which are curved posteriorly and have a serrated apical margin. Apical process of gonopods simple, blade-shaped, with pointed end. Dististylus slender. Proctiger without processes, rectangular.

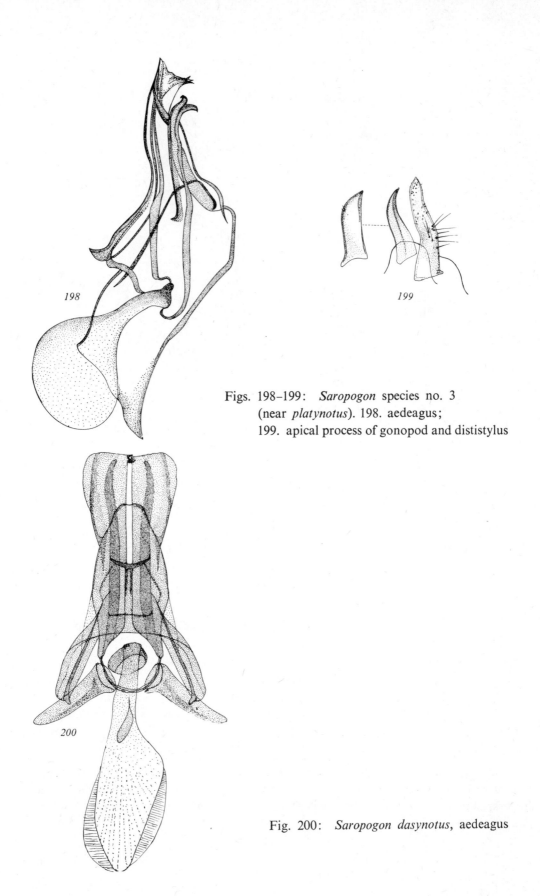

Figs. 198–199: *Saropogon* species no. 3
(near *platynotus*). 198. aedeagus;
199. apical process of gonopod and dististylus

Fig. 200: *Saropogon dasynotus*, aedeagus

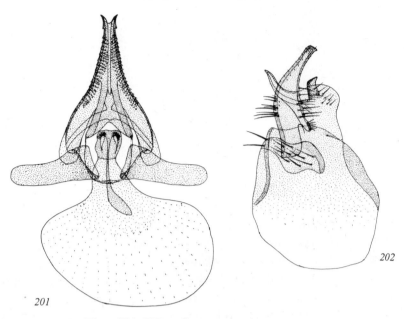

Figs. 201–202:   *Saropogon pittoproctus*
201. aedeagus; 202. apical process of gonopod and dististylus

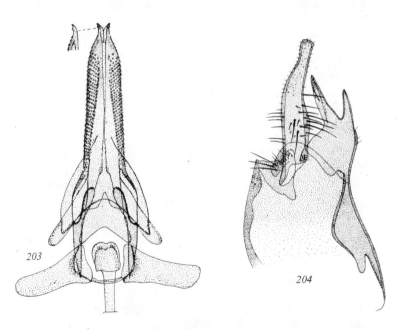

Figs. 203–204:   *Saropogon* species no. 2 (near *pittoproctus*)
203. aedeagus; 204. apical process of gonopod and dististylus

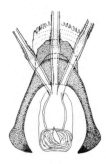

Fig. 205: *Saropogon* species no. 2 (near *pittoproctus*),
furca of spermatheca

Species no. 3 (near *platynotus*).   Aedeagus with wide, funnel-shaped end, slightly conical in the apical part. Sheath with two small, curved, horn-shaped dorsal processes in the apical part and lateral curved ridges. Apical process of gonopods simple, blade-shaped, pointed. Proctiger without processes. Dististylus slender.

dasynotus.   Aedeagus short, wide. Sheath with two large, curved processes which reach far beyond the end of the aedeagus and are curved inwards and articulated by hook-shaped processes at the end. Proctiger with short, apical lateral processes. Gonopods with simple, pointed apical process.

pittoproctus.   Aedeagus narrowly conical, with two denticles near the apex, completely covered with denticles except in a median stripe. Basal lateral apodemes very large, nearly rectangular. Proctiger with broadly rounded apical lateral processes. Apical process of gonopods with a large bulge, long, pointed end and a horn-shaped process on the surface. Dististylus curved, narrower in the apical half.

Species no. 2 (near *pittoproctus*).   Aedeagus long, narrow, nearly parallel-sided in its greater part, tapering only at the apex. A small denticle near the apex. Denticles present only laterally, bare median stripe wider. Proctiger with truncate lateral apical processes which are slightly concave apically. Apical process of gonopods bifid. Dististylus slender, less curved than in *pittoproctus*.

Species no. 4.   Aedeagus very long, slender, S-curved, tapering, without differentiations. Proctiger without processes. Apical process of gonopods with two broadly rounded, plate-shaped processes.
Spiral of spermathecae with only one or two turns, markedly thicker at the base.

Species no. 5.   Aedeagus with short apical part with two lateral denticles. An oblong area of denticles in the apical part on the ventral side. Proctiger with two apical lateral processes and two long, curved, tapering processes near the base. Apical process of gonopods trifid. Dististylus slender, slightly curved.
Spermathecae forming a spiral with 2–3 turns, markedly thicker at the base. Furca Y-shaped, i.e., angular, with a narrow apodeme.

Species no. 7.   This species has three distinct colour forms:
1. Legs reddish yellow in both sexes (Central Negev);
2. Legs black in male, reddish yellow in female (Coastal Plain);
3. Legs black in both sexes (Jordan Valley).

90

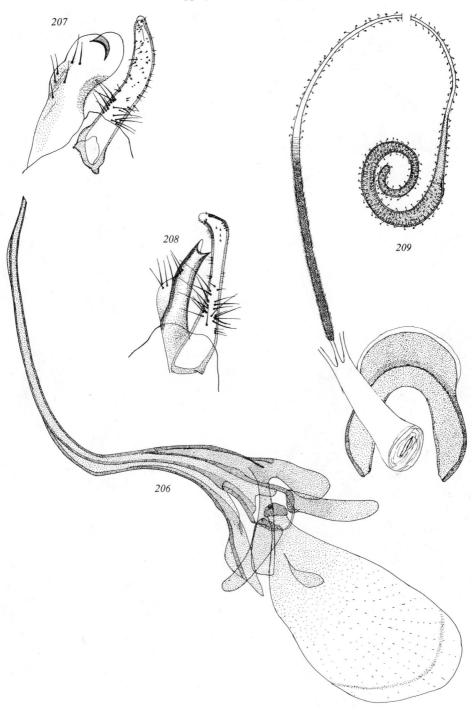

Figs. 206–209: *Saropogon* species no. 4
206. aedeagus; 207. apical process of gonopod and dististylus;
208. same, different aspect; 209. spermatheca

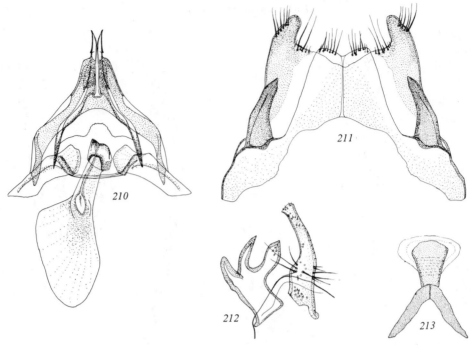

Figs. 210–213: *Saropogon* species no. 5. 210. aedeagus; 211. proctiger; 212. apical process of gonopod and dististylus; 213. furca of spermatheca

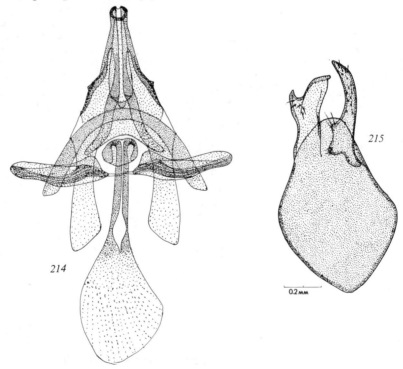

0.2 мм

Figs. 214–215: *Saropogon* species no. 7
214. aedeagus; 215. apical process of gonopod and dististylus

The genitalia of all three forms are identical. The differences in coloration are so marked that these forms would have been considered different species in the past.

Aedeagus with a broadly triangular basal and narrow, curved apical part without differentiations. Apical process of gonopods with pointed apex and short, pointed subapical process. Dististylus slender, tapering, with slightly club-shaped apex which appears pointed in lateral view.

Spermathecae forming a thick spiral with two turns. Ducts thick, with a ring-shaped sclerotization proximally, before the short, membranous posterior ducts. Furca U-shaped, with posterior lateral processes and a rounded apodeme which is narrower than the furca.

Endoaedeagus not recognizable in all the species examined above.

Molobratia teutonus (Figs. 216–217)

This species was placed in the tribe Dioctriini by Hull (1962) because of the absence of spines on the ovipositor. But the structure of the fore tibia, with an apical process and a strong spur, and a tubercle with denticles on the basitarsus on which the spur rests, resembles the structure in *Dasypogon* and *Saropogon* so closely that it seems more correct to place *Molobratia* in the Dasypogonini and to consider the absence of spines in the female genitalia as secondary.

The spermathecae form large, sclerotized spirals with 4–5 turns and thin tubes which taper to a fine end. Internal spines absent. Ducts wide, very long, with distinct openings of canaliculi. Valve not recognizable. A strongly sclerotized ring near the base. The ducts become wider posteriorly, but an ejection apparatus is not recognizable. Furca U-shaped, with narrow arms.

Aedeagus long, tubular, S-curved. Sheath with rows of denticles at the end. Endoaedeagus distinct, with spines.

Apical process of gonopods thick, bifid at the end, with a large horn-shaped process pointing in nearly the opposite direction at its base. Dististylus broad, tapering, with hooked end.

Paraphamartania syriaca

The spermathecae form long tubes which are sclerotized in the distal part. They are wider distally and taper to a fine end, forming a large, loose, flattened spiral or coil. Openings of canaliculi large, situated on tubercles. Thick part of the ducts with fine transverse striation. Valve weakly sclerotized. Ejection apparatus with fine processes, ending in a sclerotized, cylindrical part. Common duct short. Furca with U-shaped basal part and a narrower apical part, with a small, rounded apodeme.

Aedeagus conical, with two large lateral denticles in the apical third. Endoaedeagus distinct, broad, with spines.

Apical process of gonopods situated on the inner side, straight, with curved, pointed end. There is also a pointed dorsal process with fine hairs. Dististylus with broad base and a rod-shaped apical process with truncate end.

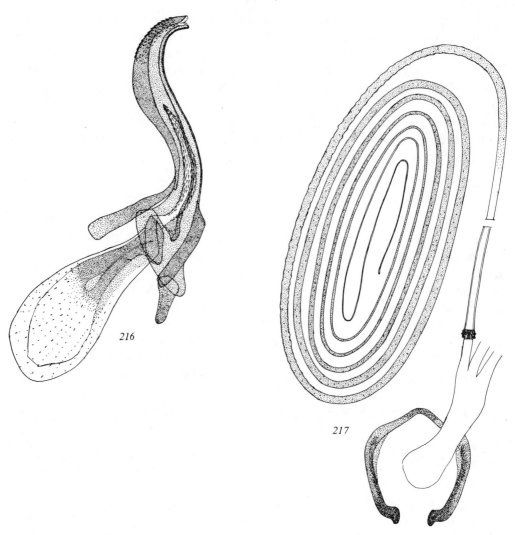

Figs. 216–217:   *Molobratia teutonus.* 216. aedeagus; 217. spermatheca

Leptarthrus brevirostris (Figs. 218–219)

The spermathecae form a large spiral with 4–5 turns. The tubes of the spiral are markedly thicker and striated in the outer turns and are covered with large canaliculi. Ducts very long, thicker, and covered with ridges and tubercles in the distal part, smooth and with inner granulations posteriorly. Valve sclerotized, cylindrical. Ejection apparatus thin-walled, striated. Furca V-shaped, with undulate outer margin and an inner process at the end of the lateral arms.

Aedeagus conical, with two lanceolate, plate-shaped, posteriorly directed processes at the apex and a ventral plate with three pointed processes. Endoaedeagus with spines. Basal plates broad, triangular. Apodeme narrow, with a wide head.

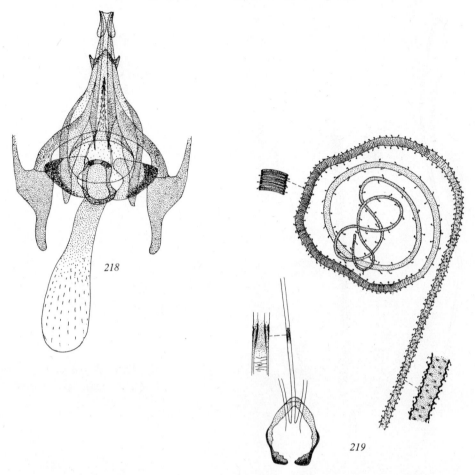

Figs. 218–219: *Leptarthrus brevirostris.* 218. spermatheca; 219. aedeagus

Neolaparus mesasiaticus and an unidentified species from South Africa (Figs. 220–221)

The spermathecae of the species from South Africa form a sclerotized, recurved tube with blunt end. Ejection apparatus very long, thin, with strong, short spines at the distal and proximal ends and with long processes its whole length. It continues in thin-walled ducts which unite in a moderately long common duct with granulations in its distal part and an inner sclerotization. Furca V-shaped, lateral arms widening posteriorly, and with a plate-shaped keel.

The aedeagus of *mesasiaticus* has a broad, nearly rectangular base in lateral view and a long, narrow, slightly S-curved, tapering apical part. Pump chamber small, apodeme narrow, with wide head, without a neck. Basal plates apparently absent.

The genera of the tribe Dasypogonini differ so markedly in the structure of the genitalia that the tribe seems heterogeneous; the character on which the tribe is based, the spine at the apex of the fore tibia, probably developed independently in different groups.

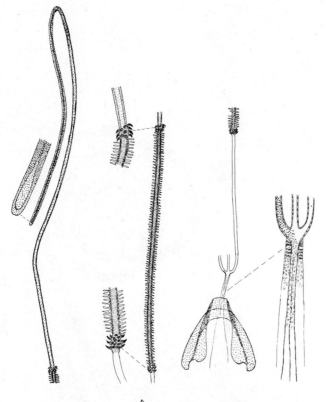

Fig. 220: *Neolaparus* sp.
(South Africa), spermatheca

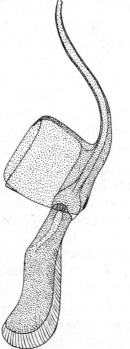

Fig. 221: *Neolaparus mesasiaticus*, aedeagus

Ommatiini

Ommatius variabilis and five unidentified species (Figs. 222–227)

The spermathecae of *variabilis* form wide, thin-walled tubes with rounded end, but they taper to a thin end in another species. The ducts are regularly striated in a zigzag pattern.

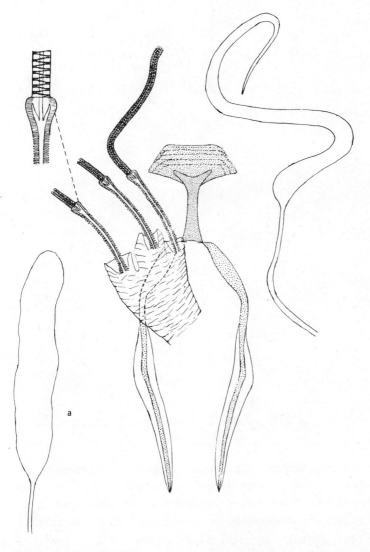

a

Fig. 222: *Ommatius variabilis*, spermatheca; (a) reservoir of spermatheca of species no. 1

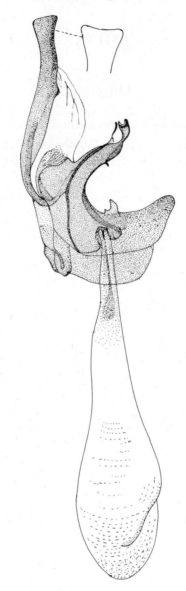

Fig. 223: *Ommatius* species no. 2, aedeagus

Valve conical, weakly sclerotized. Ejection apparatus with fine processes. The ducts end in invaginated lobes in a wide common duct. Furca complete, with thin, curved lateral arms and an apical apodeme with a narrow stem and a wide, crescent-shaped plate.

The aedeagus of four species examined shows remarkable differentiations.

Species no. 2 (Tanganyika). Aedeagus S-curved with a differentiated end and pointed denticles. There is also a large ventral process which is very long, triangular, and wider apically.

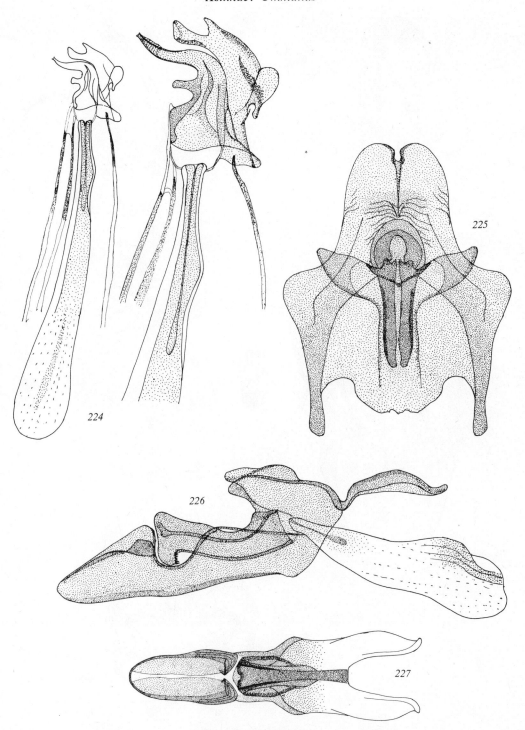

Figs. 224–227: Aedeagus of *Ommatius*. 224. species no. 3;
225. species no. 4; 226. species no. 5, lateral; 227. same, dorsal

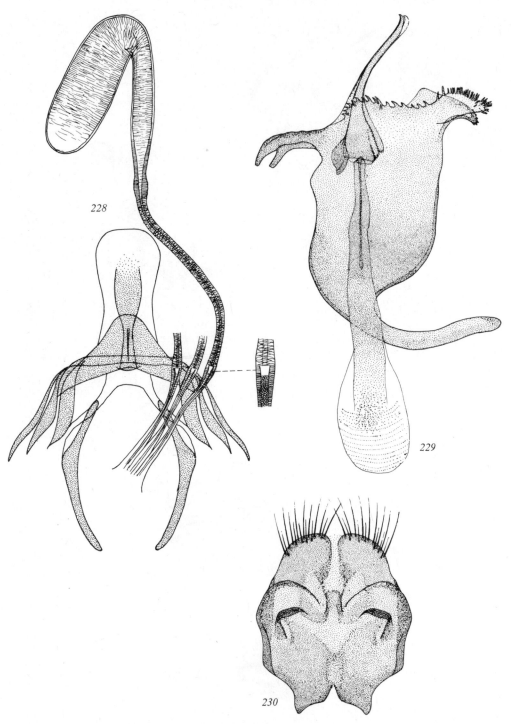

Figs. 228–230: *Cophinopoda chinensis*. 228. spermatheca;
229. aedeagus; 230. proctiger

Species no. 3 (West Africa, small). Aedeagus S-curved and ventral process of complicated form. Apodeme very long and slender. The aedeagus has long, partly sclerotized tendons instead of basal apodemes.

Species no. 4 (West Africa, large). Ventral process very wide, with rounded apex and a deep apical incision. Aedeagus short, recurved on the ventral plate.

Species no. 5 (Nigeria). Ventral process triangular in lateral view, with a deep concavity, and rounded end in dorsal view. The aedeagus enters the concavity at nearly a right angle and has denticles directed proximally at the apex.

The gonopods, their apical processes and dististylus are of distinct form in all four species.

Cophinopoda chinensis (Figs. 228–230)

The spermathecae form wide, thin-walled, striated sacs with rounded end. Valve conical, weakly sclerotized. Ejection apparatus short, striated, ending in a short common duct. Furca of complicated form, with a differentiated anterior arc, apparently divided into several arcs, and a short, broad apodeme with rounded end. The lateral arms are separated and enter the curvature of the apical part.

Aedeagus forming a long, tapering tube which is slightly bifid at the end. Sheath very wide, with two long, horn-shaped ventral processes, lateral rounded processes covered with spines, and a short dorsal process. Apodeme long and slender.

Proctiger broad, with a large, rounded bulge near the base at each side which is concave posteriorly. Gonopods short, with a slender, rod-shaped apical process. Dististylus large, broad, nearly crescent-shaped.

Asilini

Asilus crabroniformis

Spermathecae with long, relatively wide tubes with tapering end which form an irregular, loose coil. Valve conical, weakly sclerotized. Ejection apparatus very short, wide, with distinct transverse ridges. Furca complete, with wide lateral arms. Anterior part V-shaped, apodeme triangular, its wider part apical. Posterior median sclerite spindle-shaped.

Aedeagus long, curved, with wide, funnel-shaped base. Sheath narrow, closely adjacent to the pump. Prongs moderately long, of equal length. Apodeme short, narrow.

MACHIMUS GROUP

Machimus annulipes, chrysitis, fimbriatus, gonatistes, rusticus, setibarbus, tujagorum and an undescribed species
Tolmerus atricapillus, atripes, corsicus, pyragra, setiventris
Epitriptus arthriticus, cingulatus, inconstans, setulosus and four undescribed species
(Figs. 231–282)

The genera *Epitriptus* and *Tolmerus* are here considered as subgenera of *Machimus*, following Engel (1930). The division of the *Machimus* group was based in the past on external characters: number of setae on the scutellum, form of sternite 8 of the male, absence or presence of setae on the abdominal sternites, etc. Examination of the genitalia of the above

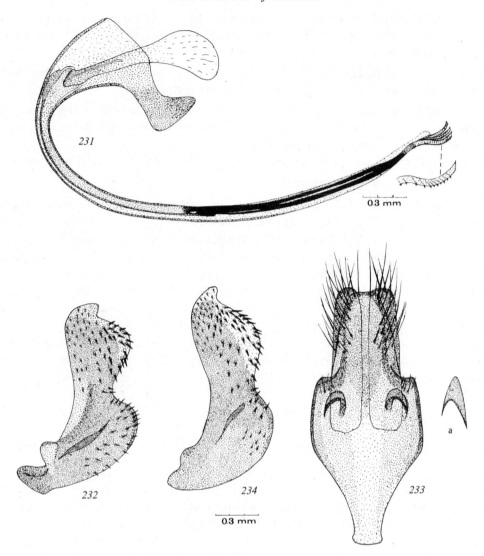

Figs. 231–233: *Machimus setibarbus* (Asia Minor). 231. aedeagus;
232. dististylus; 233. proctiger; (a) ventral process of proctiger (Israel)
Fig. 234: *Machimus setibarbus* (Israel), dististylus

21 species shows that external characters are not sufficient for the definition of groups, and that the genitalia (form of aedeagus, dististylus, gonopods, etc.) may permit a more natural grouping.

There are three types of aedeagus in the *Machimus* group, each associated with different forms of the gonopods and dististylus:

Type 1. (*Machimus* s. str.). Aedeagus long or moderately long, more or less strongly curved at the base, which is wide and funnel-shaped. Sheath narrow. Apodeme short.

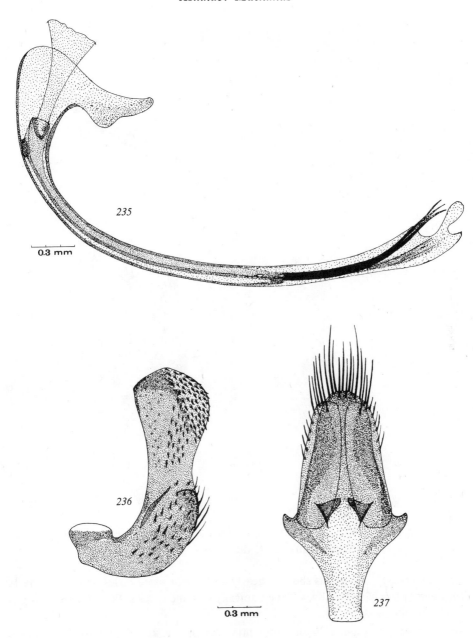

Figs. 235–237: Specimen determined as *setibarbus* from Poros
235. aedeagus; 236. dististylus; 237. proctiger

Pump nearly as long as the sheath. Prongs either very long, pigmented, or shorter and differentiated (*chrysitis*). The prongs may be of different length or curved in different directions. Gonopods short, wide, rounded-triangular. Dististylus curved at the base, short, wide, of varying form.

103

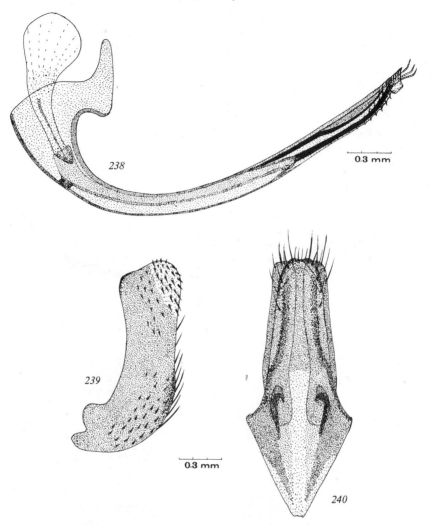

Figs. 238–240: Specimen determined as *setibarbus* from Corsica
238. aedeagus; 239. dististylus; 240. proctiger

Type 2. (*Tolmerus*). Aedeagus shorter, nearly straight or more or less curved at the base. Sheath more or less wide basally, widened apically in some species. Prongs shorter, pigmented or not. Apodeme short.

Gonopods oblong-triangular, with differentiations in the apical part on the inner side. Dististylus curved, narrow at the base, wider apically, bilobed in some species, not bilobed in others, or of different form.

Type 3. (Group 3). Three undescribed species have a different aedeagus and dististylus. Aedeagus with a wide sheath, conical, prongs short, more or less wide, slightly curved, not pigmented. Pump short, apodeme long and slender.

Gonopods oblong-triangular, with differentiations in the apical part on the inner side. Dististylus parallel-sided, slightly curved, not widening apically.

104

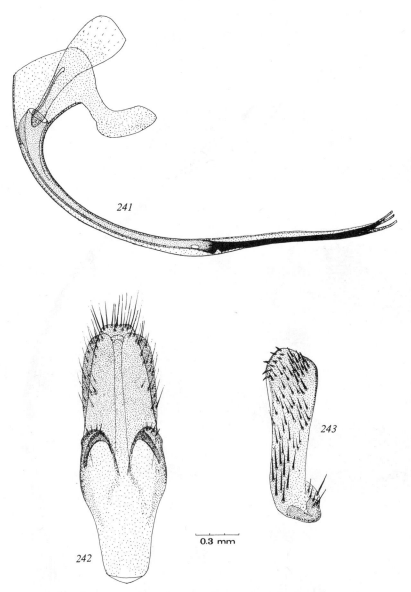

Figs. 241–243: Specimen determined as *setibarbus* from Cyprus
241. aedeagus; 242. proctiger; 243. dististylus

Machimus s. str.

The spermathecae form long tubes which are more or less sclerotized in the distal part. The three tubes together form a dense coil. End of tubes tapering. Ducts long, valve conical, weakly sclerotized. Ejection apparatus either short and wide, with processes which are longer near the valve in some species, or long and slender, with strongly sclerotized processes. Common duct long and wide. The median tube ends further posteriorly in *M. annulipes*,

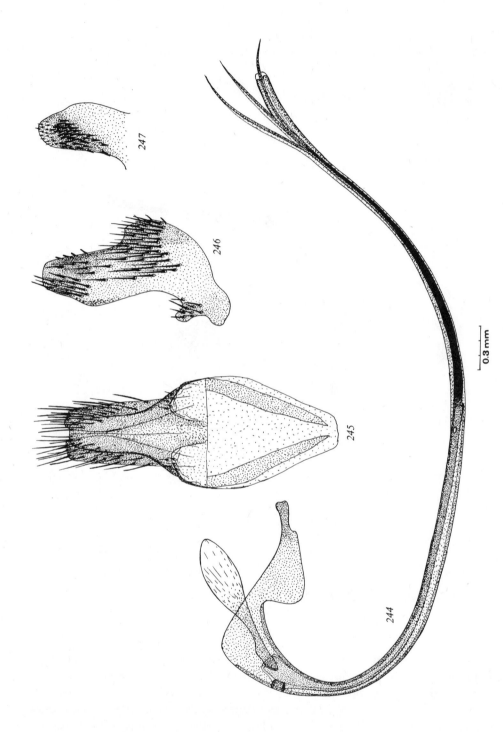

Figs. 244–247: *Machimus intermedius* from Sweden
244. aedeagus; 245. proctiger; 246. dististylus, inner side; 247. same, outer side

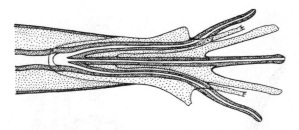

Fig. 248: *Machimus chrysitis*, end of aedeagus, flat

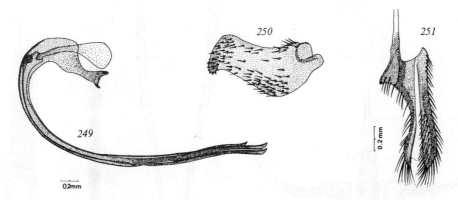

Figs. 249–251: *Machimus annulipes*. 249. aedeagus;
250. dististylus; 251. proctiger, lateral

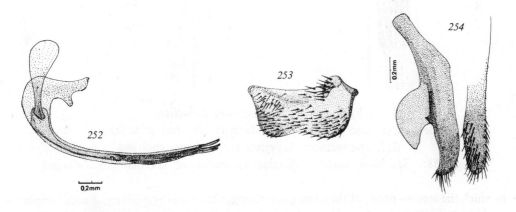

Figs. 252–254: *Machimus rusticus*. 252. aedeagus;
253. dististylus; 254. proctiger, lateral

107

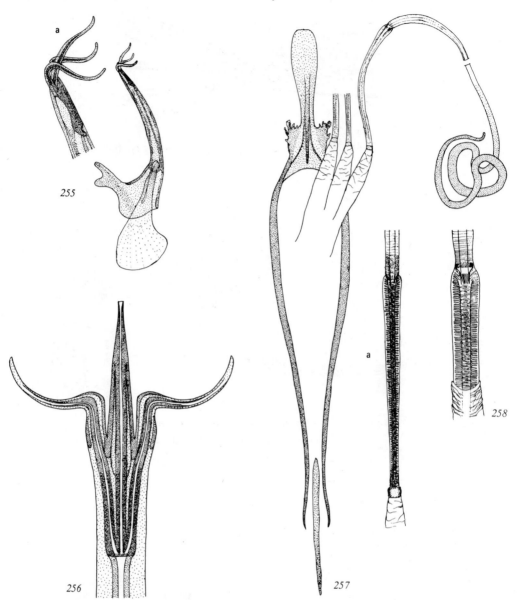

Figs. 255–257: *Machimus gonatistes*
255. aedeagus; (a) end of aedeagus, enlarged; 256. end of aedeagus, dorsal;
257. spermatheca; (a) ejection apparatus, enlarged
Fig. 258: *Machimus rusticus*, ejection apparatus of spermatheca, enlarged

in which the median prong of the aedeagus is shorter than the lateral prongs. Furca complete, moderately narrow, apodeme long. Median posterior sclerite long, rod-shaped. There are differences in the width of the tubes, their degree of sclerotization and the proportions of the various parts in the different species.

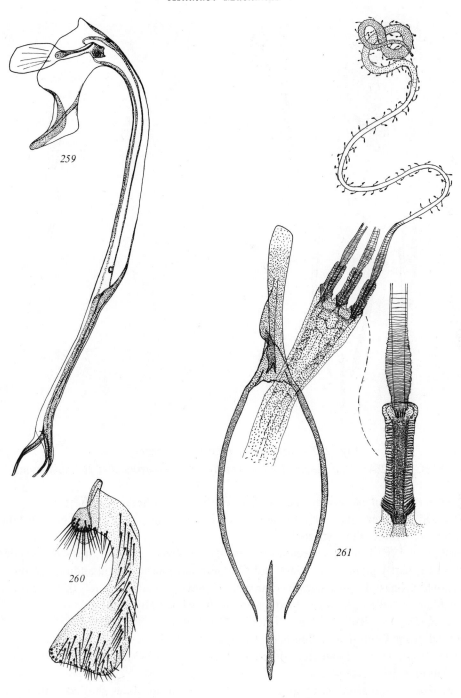

Figs. 259–261: *Machimus* species no. 1. 259. aedeagus;
260. dististylus; 261. spermatheca

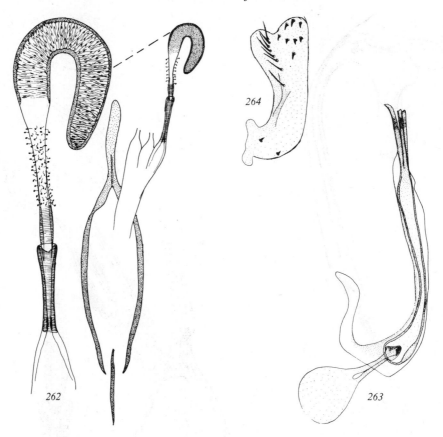

Fig. 262: *Machimus tujagorum*, spermatheca

Figs. 263–264: *Machimus (Tolmerus) pyragra*. 263. aedeagus; 264. dististylus

M. tujagorum is an exception. The reservoir of the spermathecae is a short, wide, strongly sclerotized, recurved tube with reticulate structure. Ducts very short. Valve broadly conical. Ejection apparatus short, striated.

Aedeagus of Type 1 in most species. *M. chrysitis* has a shorter, wider, less curved aedeagus with short, partly pigmented prongs and four secondary processes in addition to the three functional prongs. *M. gonatistes* has three short, not pigmented prongs, the lateral prongs curved laterally at a right angle. Median prong curved dorsally in species no. 1.

M. setibarbus was described from 'Rhodes, Greek islands and Asia Minor'. It was later recorded from Central and Southern Europe by Engel. Lyneborg (1968) synonymized *Machimus intermedius* Holmgren, described from Sweden, with *setibarbus*. *M. setibarbus* is common in Israel.

Comparison of the genitalia of specimens from different localities showed that specimens from Israel and Trieste closely resemble specimens from Asia Minor (Uludagh, near Brussa). However, specimens from Corsica, Poros (Greece) and Cyprus showed distinct differences in the form of the aedeagus and dististylus. The aedeagus of the forms from Poros and Corsica showed specific modifications of the apical part which are absent in *setibarbus*, and

110

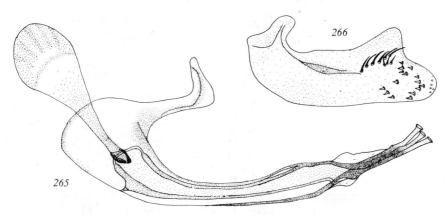

Figs. 265–266: *Machimus (Tolmerus) atricapillus*
265. aedeagus; 266. dististylus

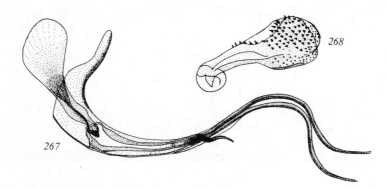

Figs. 267–268: *Machimus corsicus.* 267. aedeagus; 268. dististylus

the dististylus is of different form. The aedeagus of a specimen from Cyprus has shorter prongs and the end of the prongs is straight and without the fine ridges present in *setibarbus*. The dististylus is also of different form. This specimen also differs from *setibarbus* in the straight posterior margin of sternite 8 and in the partly reddish tibiae.

Examination of a specimen of '*intermedius*' from Sweden (kindly sent by Dr L. Lyneborg, Copenhagen) showed that it also differs distinctly from *setibarbus*. Sternite 8 is less produced posteriorly and the arrangement of the setae is different. The epandrium is shorter and more rounded apically. The dististylus is of different form and has a large group of short, black spines on the outer side of the apical part. The aedeagus is very long, much longer than in *setibarbus*, with a long, pointed end of the prongs and a ventral, plate-shaped structure below the prongs.

Thus there are apparently several species which closely resemble *setibarbus* in external characters, but which show distinct differences in the genitalia.

Tolmerus, Epitriptus

T. pyragra. Aedeagus short, slightly curved at the base. Sheath wide at the base, narrowing and then widening again before the prongs, which are moderately long and pigmented.

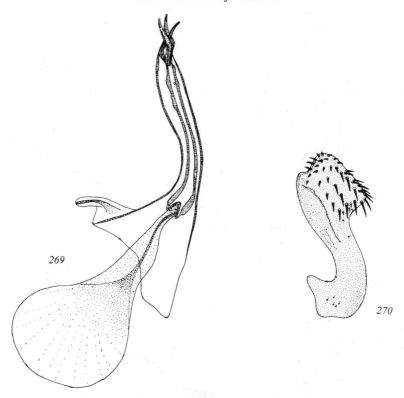

Figs. 269–270: *Machimus setulosus*. 269. aedeagus; 270. dististylus

Gonopods oblong-triangular, with three curved ridges at the apex on the inner side. Dististylus curved at the base, wider and bilobed apically.

Spermathecae with weakly sclerotized tubes which form a loose coil. Ejection apparatus short, striated. Furca relatively wide, complete.

T. atricapillus. Aedeagus, gonopods and dististylus as in *pyragra*, with minor differences.

Spermathecae with long, delicate tubes which do not form a coil. Ducts wider posteriorly, valve broadly conical, with internal spines. Ejection apparatus short, striated, with an internal sclerotization posteriorly.

The genitalia resemble those of *T. pyragra* so closely that *atricapillus* is considered here to belong to *Tolmerus* in spite of the long, produced, bifid sternite 8 of the male.

T. atripes. Aedeagus as in *T. pyragra*, but shorter, more strongly curved at the base, prongs markedly longer, as long as the narrow part of the sheath. Gonopods oblong-triangular, with a network of ridges with spines near the apex on the inner side. Dististylus bilobed.

Spermathecae with narrow tubes which form a loose coil. Ducts wider, otherwise as in *pyragra*.

Species no. 2. (*Tolmerus*). Aedeagus as in *T. pyragra*, but prongs shorter, not pigmented. Gonopods broad, with an apical point and three ridges at the apex on the inner side. Dististylus deeply bilobed.

112

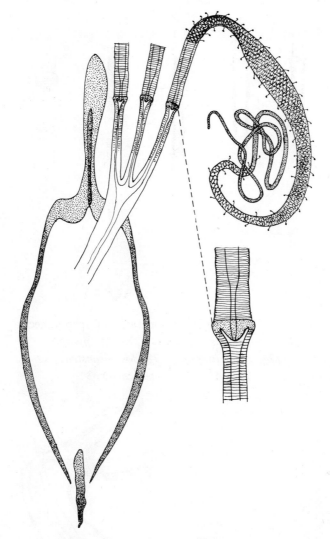

Fig. 271: *Machimus* species no. 2, spermatheca

Tubes of spermathecae narrow in the distal part, forming a loose coil. They continue in a wide part with reticulate structure. Valve distinct, broadly conical. Ejection apparatus moderately long, narrow, striated.

M. arthriticus. Aedeagus of Type 1. Long, strongly curved at the base. Sheath narrow, prongs pigmented, long. Gonopods broadly rounded, with a small, pointed process in the middle of the apical margin. Dististylus curved at the base, very short and wide, with angular apex. This species belongs to *Machimus* s. str. according to the male genitalia.

T. corsicus. Aedeagus slightly curved, short, sheath moderately wide, tapering apically. Lateral prongs very long, S-curved, about six times as long as the median prong, which is curved ventrally.

Gonopods parallel-sided in basal part, broadly rounded apically, with 5–6 parallel, trans-

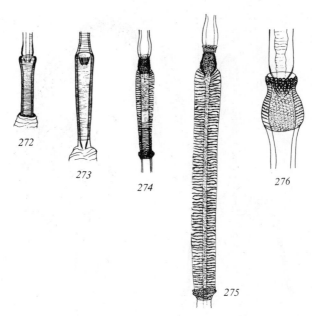

Figs. 272–276: Ejection apparatus of spermatheca
272. *Machimus cingulatus*; 273. *M. pyragra*;
274. *M. setiventris*; 275. *M. inconstans*;
276. *Machimus* species no. 3

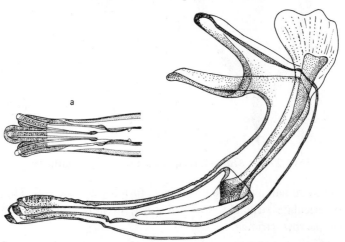

Fig. 277: *Machimus* species no. 3, aedeagus; (a) end of aedeagus, dorsal

verse ridges in the apical part on the inner side. Dististylus wider apically but not bilobed, with a plate-shaped process at the apex.

E. cingulatus. Aedeagus short, slightly curved, sheath nearly tubular. Prongs short, straight, not pigmented, wide. Dististylus widening apically, apical concavity very shallow. Gonopods broad, triangular, with pointed apex and 3–4 ridges in the apical part on the inner side.

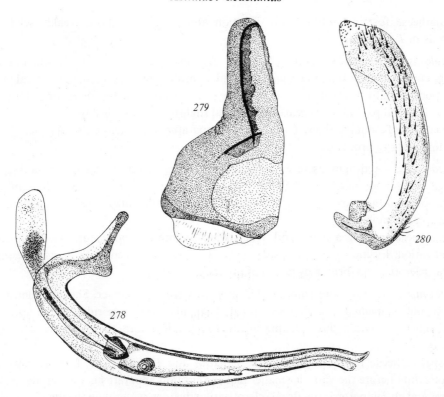

Figs. 278–280: *Machimus* species no. 5. 278. aedeagus;
279. gonopod; 280. dististylus

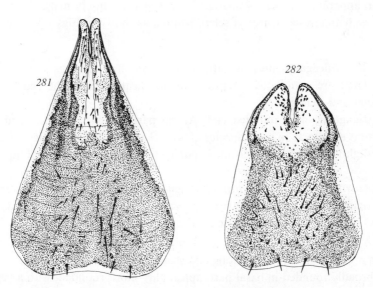

Figs. 281–282: Sternite 8 of female
281. *Machimus* species no. 5; 282. *Machimus* species no. 4

115

Spermathecae forming a loose coil. Ejection apparatus very short, weakly sclerotized. Furca as in *T. pyragra*.

E. setulosus.   Aedeagus resembling that of *E. cingulatus*, but shorter, sheath wider apically. Prongs curved, shorter than in *cingulatus*, pointed, median prong crossing the lateral prongs. Gonopods broadly triangular, with one ridge at the apex on the inner side. Dististylus with wide apical part which is nearly square and slightly bilobed apically.

Spermathecae forming a dense coil. Ejection apparatus long, slender, strongly sclerotized, with long, strong processes.

E. inconstans.   Aedeagus as in *E. setulosus*, but prongs shorter and wider, pointed, not crossed. Gonopods narrower than in *E. cingulatus*, with S-curved dorsal margin and a ventral point at the apex. One ridge in the apical part on the inner side. Dististylus slender, wider apically, not bilobed.

Spermathecae forming a dense coil. Ducts wider before the valve, forming a cup-shaped sclerotization. Ejection apparatus long, wide, strongly sclerotized, with long processes and a basal sclerotization formed by fusion of processes.

T. setiventris.   Aedeagus resembling that of *T. pyragra*, but shorter. Sheath wider, prongs shorter, not pigmented, tapering. Gonopods triangular, with one ridge in the apical part on the inner side. Dististylus widening slightly to the apical third, then tapering to a truncate end.

The spermathecae resemble those of *E. setulosus* and *E. inconstans*, but the cup-shaped sclerotization before the valve is smaller and the ejection apparatus is shorter, its processes shorter and absent posteriorly. Basal sclerotization distinct, wider than the duct.

T. cyrnaeus.   Dististylus bilobed as in *T. pyragra*, according to the drawing given by Oldroyd (1946).

The ejection apparatus of the spermathecae of *Tolmerus* differs markedly in the various species in length and in the degree of sclerotization, as shown in Figs. 272–276.

Group 3

Species no. 3.   Aedeagus conical, sheath wide, pump short. Prongs short and wide, not pigmented. There are granulated extensions of the sheath around the end of each prong. Apodeme long and slender.

Gonopods triangular, with rounded end. Apical part narrow, gutter-shaped, with dark, serrated ventral margin as in some species of *Acanthopleura* and in species no. 5 (Fig. 279). Dististylus slightly curved, parallel-sided, not widening apically, with obliquely truncate end.

Spermathecae with delicate tubes forming a dense coil. Ejection apparatus very short and wide, bulb-shaped with reticulate structure. Valve indistinct. Furca divided, apical part very narrow, lateral arms reaching close to the apical part.

Species no. 4.   Aedeagus as in species no. 3, but shorter and wider. Prongs short, narrow, curved. Pump very short. Apodeme long and slender.

Gonopods broadly rounded in basal part, apical part short, rounded, with a ventral corner and a curved ridge on the inner side in the apical part, as in some species of *Tolmerus*. Dististylus parallel-sided, with rounded end and dense, small tubercles.

Spermathecae with very long, narrow tubes which form a loose coil. Ejection apparatus short, wide, striated.

Species no. 5.   Aedeagus as in species no. 4, but prongs longer and wider, pump longer.
Gonopods broad at the base, with a narrow apical process which is gutter-shaped, and a dark, serrated ventral margin as in species no. 3. Dististylus as in species no. 4, but with pointed end and without tubercles.
Spermathecae with long, delicate tubes, wider than in species no. 4. Valve with internal spines.
Sternite 8 of female of markedly different form in species nos. 4 and 5 (Figs. 281–282).

The following conclusions may be drawn:
1. *M. atricapillus* resembles *T. pyragra* so closely in the form of aedeagus and dististylus that it obviously belongs to *Tolmerus* in spite of the long, produced, bifid sternite 8 of the male.
2. *setiventris* also resembles *Tolmerus* in the genitalia, but it also has a long, triangularly produced sternite 8 in the male.
3. The presence of setae on the abdominal sternites was considered a generic character of *Epitriptus* by Hull (1962). They are present in *E. arthriticus, inconstans* and *setulosus*, but *cingulatus*, the type species of *Epitriptus*, has only long, thin hairs on the abdominal sternites.
4. *Epitriptus* and *Tolmerus* are apparently indistinguishable, and *Epitriptus* should be considered a synonym of *Tolmerus*.
5. The species of group 3 resemble *Tolmerus* in external characters, but differ so markedly in the genitalia that they may have to be placed in a separate subgenus. Gonopods and dististylus of species nos. 3 and 5 resemble those of some species of *Acanthopleura*.
6. The above grouping is provisional and may have to be changed if further species are examined.
7. The genitalia give distinct characters to distinguish closely related species like *M. annulipes, rusticus* and *gonatistes*, as described by Weinberg (1961, 1967). *M. gonatistes* was considered a synonym of *rusticus* by Engel (1930) and Hull (1962).
8. The spermathecae of *M. annulipes* and *rusticus* are similar, except for minor differences and in that the tubes of the reservoir of *rusticus* are weakly sclerotized. The ejection apparatus is short and wide and weakly sclerotized. The ejection apparatus of *gonatistes* is very long and slender, with strongly sclerotized processes which are fused in a network at the base. The furca is much longer than in the other two species and has distinct processes with a ragged margin at the base of the apodeme.
9. The prongs of the aedeagus of *M. rusticus* are of the same length, the median prong is shorter than the lateral prongs in *M. annulipes*, and the lateral prongs are curved laterally at a right angle in *M. gonatistes*. The gonopods and dististylus are of different form in the three species and the epandrium of *gonatistes* is nearly rectangular, not slender, and with a rounded end as in the other two species.
10. A character not mentioned by Weinberg is the form of the proctiger. This has large, plate-shaped ventral processes in *M. rusticus*, small, pointed processes in *M. annulipes*, and no processes in *M. gonatistes*.

Acanthopleura brunnipes, goedli and an undescribed species (Figs. 283–287)

The aedeagus, gonopods and dististylus of *brunnipes* differ markedly from those of *goedli*, of the new species and of other species of *Acanthopleura* illustrated by Tsacas (1964, 1967).

brunnipes. Aedeagus of the *Machimus* type, curved, with a narrow sheath and long, pigmented prongs. Sheath slightly wider in the apical part. Gonopods short, broadly rounded-triangular. Dististylus short, broad, with an apical corner.

The spermathecae form a coil with wide, strongly sclerotized tubes of reticulate structure. Valve, ejection apparatus and furca as in species of *Machimus*.

goedli. Aedeagus conical, sheath wide, tapering apically. Pump short. Prongs wide, slightly curved, not pigmented. Apodeme long and slender.

Gonopods broad at the base, with a long apical process which is gutter-shaped and has a serrated ventral margin. Dististylus long, slender, parallel-sided, slightly curved.

Spermathecae with wide, weakly sclerotized tubes which form a loose coil. Ejection apparatus short, with distinct processes.

Acanthopleura sp. Aedeagus as in *goedli*, but shorter, prongs also shorter. Gonopods and dististylus also as in *goedli*, with minor differences.

Spermathecae with sclerotized tubes which form a dense coil. Ducts wide, sclerotized and with internal spines before the conical valve. Ejection apparatus short, with distinct processes. Furca long, narrow, posterior median sclerite long, rod-shaped.

The aedeagus of *A. keiseri*, as illustrated by Tsacas (1967), resembles that of *goedli*. The genitalia of *goedli* and of the new species resemble those of species of group 3 of *Machimus* and of *Eutolmus facialis*.

The status of the genus *Acanthopleura* is doubtful. It has been based mainly on the presence of a row of setae near the dorsal margin of the mesopleura. But setae on the mesopleura are present also in some species of *Machimus* (*annulipes, arthriticus*). The new species of *Acanthopleura* and species nos. 3 and 5 of group 3 of *Machimus* have only a single seta on the mesopleuron. It seems doubtful, therefore, whether *Acanthopleura* should be maintained as a genus. *brunnipes* could be placed in *Machimus* s. str., according to the genitalia, and the other species with different genitalia may form a subgenus of *Machimus*, perhaps together with group 3 of *Machimus*.

Eutolmus facialis, rufibarbis and an undescribed species (Figs. 288–289)

rufibarbis. The reservoir of the spermathecae forms short, wide, strongly sclerotized tubes which are irregularly S-curved. Ducts moderately long. Valve conical. Ejection apparatus short, wide, striated. Furca complete, nearly elliptical, lateral arms wide. Apodeme short, wide, with a narrower base. Posterior median sclerite curved, thicker posteriorly.

Aedeagus of *Machimus* type, long, strongly curved, prongs long, pigmented. Pump long, apodeme short, triangular.

Gonopods short, rounded-triangular. Dististylus slightly curved, parallel-sided, moderately wide.

Eutolmus sp. The spermathecae form thick, sclerotized tubes which are narrower than in *rufibarbis* and form a loose coil. Ducts long, wider posteriorly. Ejection apparatus short, wider than the ducts, striated. Median duct ending further posteriorly than the lateral

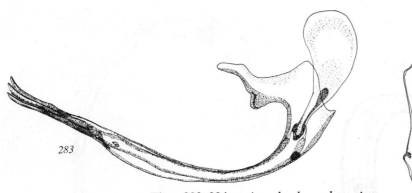

Figs. 283–284: *Acanthopleura brunnipes*
283. aedeagus; 284. gonopod and dististylus

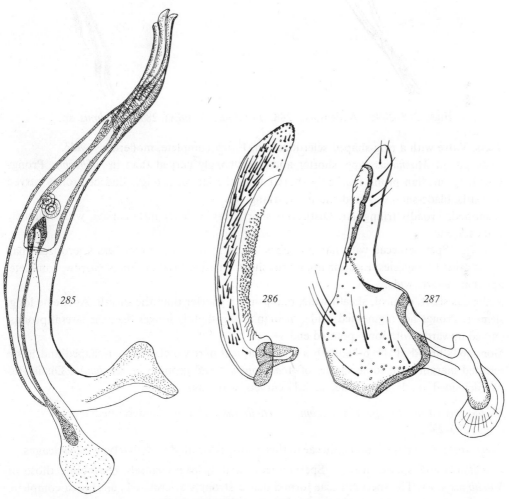

Figs. 285–287: *Acanthopleura* sp.
285. aedeagus; 286. dististylus; 287. gonopod

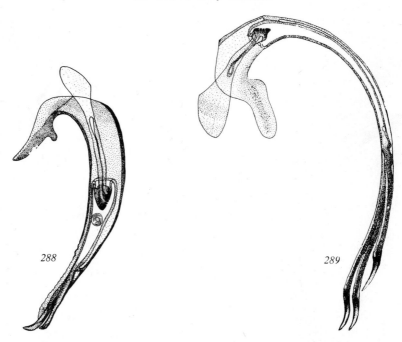

Figs. 288–289: Aedeagus. 288. *Eutolmus facialis*; 289. *Eutolmus* sp.

ducts. Valve with a ring-shaped sclerotization. Furca complete, moderately wide.

Aedeagus of *Machimus* type, shorter and less strongly curved than in *rufibarbis*. Prongs very long, median prong markedly shorter than the lateral prongs. End of prongs curved upwards, blade-shaped. Apodeme short, triangular.

Gonopods broadly triangular. Dististylus short, very wide, slightly curved, with obliquely truncate end.

facialis. Spermathecae with narrow tubes which form a loose coil, less sclerotized than in the other two species. Ejection apparatus short, narrow, striated. Furca narrow, complete, apodeme short, wide.

Aedeagus short, conical, sheath wide, pump much shorter than the sheath. Apodeme long, slender. Prongs not pigmented, wide, median prong slightly longer than the lateral prongs, curved upwards, with long, pointed end.

Gonopods broad at the base, with a narrow apical part which is gutter-shaped and has a serrated margin as in some species of *Acanthopleura* and group 3 of *Machimus*. Dististylus parallel-sided, moderately wide, slightly curved, with rounded end.

Dysmachus cochleatus, picipes, trigonus, verticillatus and three undescribed species (Figs. 290–297)

There are three types of spermathecae in this genus, associated with two types of aedeagus.

verticillatus and species no. 1. Spermathecae and aedeagus closely resembling those of *Machimus* s. str. The spermathecae form a dense, strongly sclerotized coil. Furca complete. Aedeagus long, curved at the base, with a narrow sheath and long, pigmented prongs. Gonopods resembling those of some species of *Tolmerus*, with one ridge at the apex on

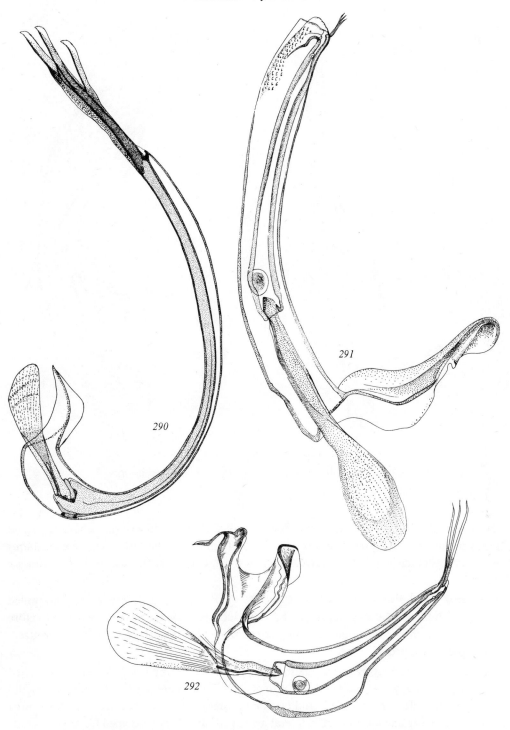

Figs. 290–292: Aedeagus of *Dysmachus*. 290. *D. verticillatus*;
291. *D. picipes*; 292. *D. cochleatus*

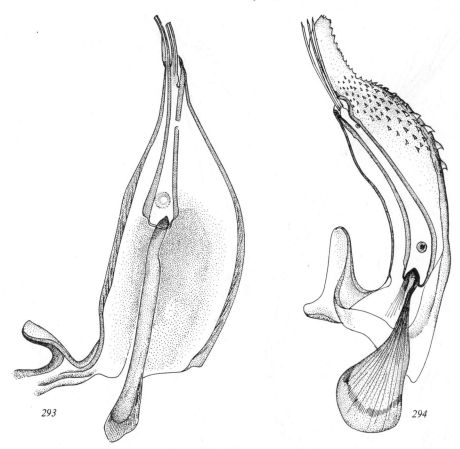

Figs. 293–294: Aedeagus
293. *Dysmachus trigonus*; 294. *Dysmachus* species no. 2

the inner side. The two species show only small differences in external characters, and differ mainly in the presence of a distinct row of spines on the dististylus of *verticillatus* which is absent in the other species. Dististylus short, curved at the base, wide and truncate apically.

cochleatus, picipes and species nos. 2 and 3. The spermathecae are wide, thin-walled tubes which taper to a long, recurved end. The ducts become narrower and striated before the valve, which is conical and strongly sclerotized. Ejection apparatus long, sclerotized, with processes. Furca divided.

trigonus. The spermathecae are again different. The reservoir is a very wide, thin-walled sac. The sacs narrow into narrow ducts with processes which are partly sclerotized internally and become wider, thick-walled, and crinkled posteriorly. Valve not recognizable. Furca divided, with a triangular apodeme, lateral arms with broad, plate-shaped inner extensions. The spermathecae of species no. 2 show a special differentiation. They form wide, thin-walled tubes with a recurved end. The ducts widen before the valve into a globular bulb

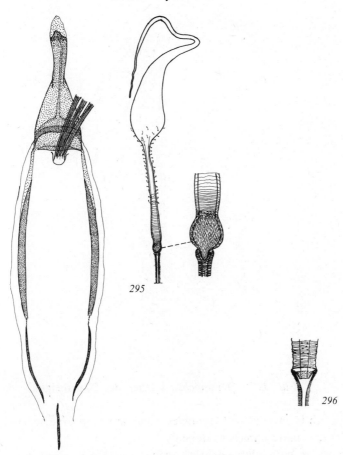

Fig. 295:   *Dysmachus* species no. 2, spermatheca
Fig. 296:   *Dysmachus picipes*, valve of spermatheca

with internal spines. The bulb opens with a narrow funnel into the conical valve. Furca of characteristic form, apical part with straight posterior margin, with a short, triangular apodeme. Furca of species no. 3 similar, but apical part deeply incised posteriorly, and a bulb before the valve is absent.

The aedeagus of all these species is transitional to the *Neomochtherus* type, with rudimentary prongs. It is more or less straight, with a wide sheath, and the prongs are reduced but functional in some species, not functional in *picipes* and species no. 3.

trigonus.   Sheath of aedeagus very wide at the base, laterally compressed, convex dorsally and ventrally. Aedeagus pump short, apodeme slender, as long as the pump. Prongs short, wide, partly covered by the sheath.

cochleatus.   Aedeagus similar, but the sheath is narrower, conical, with a ventral bulge in the basal part. Prongs much longer, tapering, thin, but apparently functional. Gonopods with long apical process, dististylus slender.

Species no. 2.   Sheath of aedeagus more or less tubular, wider distally, extending in a narrow process beneath the prongs, which are narrow and functional. The distal ventral

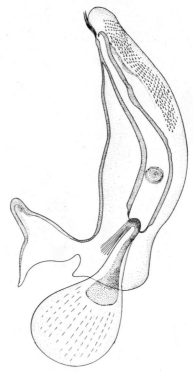

Fig. 297: *Dysmachus* species no. 3, aedeagus

bulge of the sheath is covered with denticles which are very large ventrally and gradually become smaller and more numerous dorsally.

Gonopods triangular, with a large denticle on the ventral margin and with a slender, pointed apex. Dististylus slightly wider apically.

Species no. 3.   Sheath wide, tapering apically, convex ventrally, extending in a process beneath the prongs, which are non-functional. Ventral side with rows of denticles in the distal part which become smaller apically. Pump very long, wide, curved.

Gonopods triangular, with a posteriorly directed denticle on the inner side. Dististylus slender, slightly curved. Sternite 8 with three lobes with tufts of long setae.

picipes.   Sheath tubular, truncate at the distal end. Pump very long, prongs very short, non-functional. Ventral side of sheath with fine, blunt denticles in the distal part.

Gonopods with a long apical process with transverse ridges and ventral apical point. Dististylus very slender, nearly straight.

Engel (1930) recognized a group of species in *Dysmachus* (*verticillatus, decipiens* and *periscelis*) in which the cerci of the female are not wedged into tergite 9 as in most other species of *Dysmachus* and in *Eutolmus*, but are triangular, pointed, and diverging; but he did not name this group. The differences in the male genitalia and spermathecae between the *verticillatus* group and the other species support his view and suggest that the *verticillatus* group should be considered at least a subgenus. According to other characters *decipiens* too probably does not belong to the genus *Dysmachus*.

124

Antiphrisson adpressus, trifarius (Figs. 298–299)

Spermathecae with delicate tubes which form a coil. The ducts widen markedly proximally and have a distinct reticulate structure in *adpressus;* they then narrow again before the conical valve. Ejection apparatus short, striated. Spermathecae of *trifarius* similar, but the ducts are not widened and are without reticulate structure. Furca complete, relatively short. Apodeme widening apically. Posterior median sclerite long, rod-shaped in *trifarius*, markedly reduced in one population of *adpressus* (Beersheba), absent in another (Jerusalem). Aedeagus of the *Tolmerus* type, nearly straight, with slightly widened sheath and long prongs which are less than half as long as the tubular basal part in *trifarius* but half as long or longer in *adpressus*. Apex of dististylus rounded in *adpressus*, longer and pointed in *trifarius*.

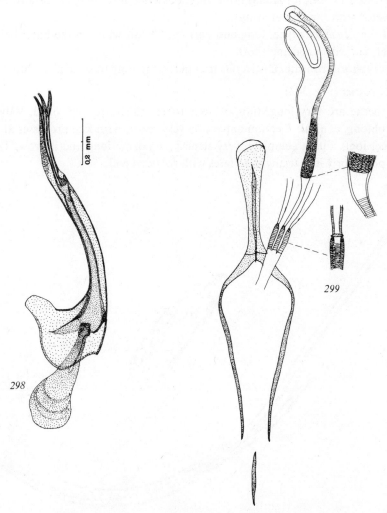

Figs. 298–299: *Antiphrisson adpressus*. 298. aedeagus; 299. spermatheca

Rhadiurgus variabilis (Figs. 300–301)

The spermathecae are wide, sclerotized, wrinkled tubes with several irregular bends. The ducts are shorter than the reservoir. Valve conical, weakly sclerotized. Ejection apparatus short, wide, striated. The two lateral ducts open markedly posterior to the median duct, corresponding to the form of the aedeagus. Furca complete, broad, apodeme long, wide.

Aedeagus of the *Machimus* type, not very wide at the base, curved. The two lateral prongs are short, the median prong much longer, visible in pinned specimens above the epandrium. Gonopods triangular, with rounded end. Dististylus short, straight, parrallel-sided in its greater part, tapering to a slightly curved, narrow end with a spine at the apex.

Astochia virgatipes

The spermathecae form irregularly coiled, sclerotized tubes with reticulate structure. Valve and ejection apparatus weakly sclerotized. Furca very slender, complete. Apodeme very long and narrow. End of lateral arms bifid. Posterior median sclerite short, triangular, partly connected with the lateral arms.

Aedeagus of the *Machimus* type, long and narrow. Sheath wider in the basal half. Prongs short, slender, curved. Apodeme short.

Gonopods very short, truncate. Dististylus triangular, tapering to a point, slightly curved.

Neoitamus cyanurus (Fig. 302)

The spermathecae are very long, thin, delicate tubes which are not coiled. Valve weakly sclerotized, oblong, conical. Ejection apparatus very short, with fine processes at the base. Common duct long. Furca complete, very slender, apodeme long and narrow. The lateral arms widen posteriorly into triangular plates with rounded end.

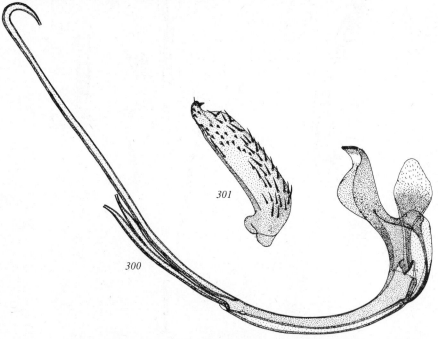

Figs. 300–301: *Rhadiurgus variabilis*.   300. aedeagus; 301. dististylus

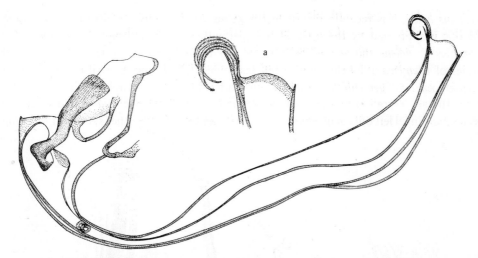

Fig. 302: *Neoitamus cyanurus*, aedeagus; (a) end of aedeagus, enlarged

Aedeagus curved and with a narrow sheath in the basal half, but widening distally to a wide, truncate end with a single opening. Prongs reduced to two tapering, curved, granulated processes. The sheath extends above the prongs in a pointed process.
Gonopods short, triangular. Dististylus crescent-shaped, wide.

Neomochtherus flavipes, illustris, macropygus, pallipes (Figs. 303–306)

The spermathecae form wide, thin-walled sacs with a more or less long, tapering end. Ducts narrow, widening before the valve in *illustris*, but narrowing again in *macropygus*. Valve strongly sclerotized, oblong-conical, with minor differences in the various species. Ejection apparatus long, narrow, with sclerotized processes which are longer and thicker proximally. Furca divided.
Aedeagus with a wide sheath, slightly curved, with a single opening. Prongs rudimentary, originating on the dorsal wall of the pump. Sheath of *illustris* nearly tubular, with relatively few denticles in the distal part. Pump long. Aedeagus shorter, slightly conical, and sheath extending further distally in *macropygus*, denticles larger and more numerous. The sheath is wider in the middle of the ventral side in *flavipes* and has a bulge in the distal part on the ventral side in *pallipes*.
Gonopods of varying form, broadly rounded in *flavipes*, very long in *macropygus*. Dististylus long, slender, slightly curved, with specific differences. It is longer than the gonopod, straight, tapering to a narrow, rounded end in *macropygus*.

Cerdistus erythrurus, geniculatus, (?)oblitus, syriacus and two undescribed species
(Figs. 307–314)

The genus is closely related to *Neomochtherus*. Some species have been placed in *Neomochtherus* which seem to belong to *Cerdistus* according to the genitalia. The definition of *Cerdistus* on external characters is clearly unsatisfactory. There are several types of spermathecae and aedeagus in the species examined. *Machimus syriacus* resembles *C. erythrurus*, the type species of the genus, in the form of the aedeagus and of the valve of the spermathecae, but it differs in the form of the reservoir.

Cerdistus heleni was recently placed in the genus *Eremisca* by Abbassian-Lintzen (1964), and this is supported by the form of the spermathecae and aedeagus. *C. pallidus* closely resembles *C. heleni* and also apparently belongs to *Eremisca*. *C. osiris* differs from other species of *Eremisca* in its chaetotaxy, but resembles them in the form of the aedeagus and spermathecae. *C. geniculatus* differs from both the above groups in the form of the spermathecae. Verrall (1909) created the genus *Paritamus* for *geniculatus* and three other species, but Engel left only *geniculatus* in *Cerdistus* and placed the other species in *Machimus*.

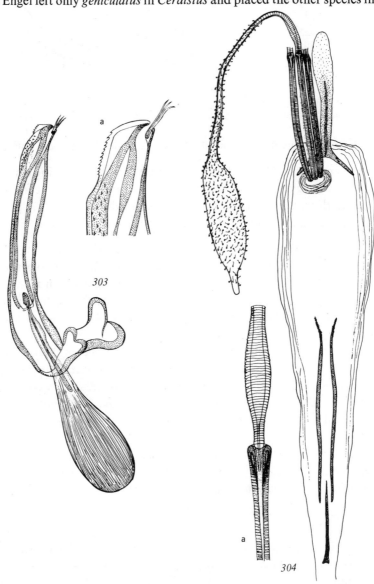

Figs. 303–304: *Neomochtherus illustris*
303. aedeagus; (a) end of aedeagus, enlarged;
304. spermatheca; (a) valve of spermatheca

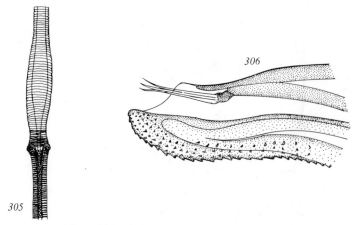

Figs. 305–306: *Neomochtherus macropygus*
305. valve of spermatheca; 306. end of aedeagus

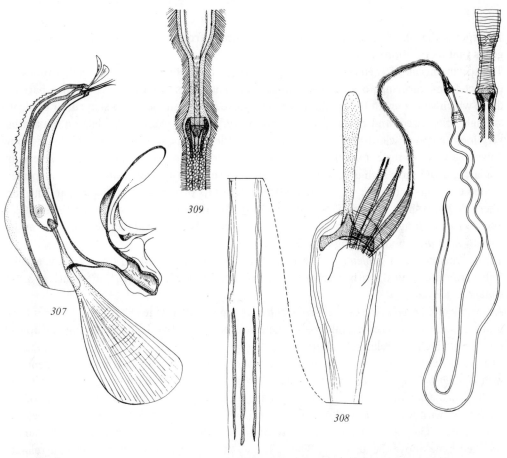

Figs. 307–308: *Cerdistus erythrurus.* 307. aedeagus; 308. spermatheca
Fig. 309: *Cerdistus* sp., valve of spermatheca

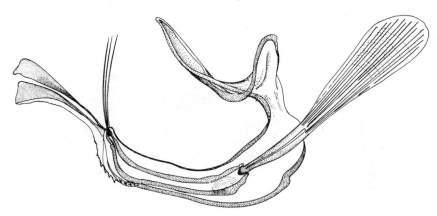

Fig. 310: *Cerdistus* (?) *oblitus*, aedeagus

The spermathecae of *geniculatus* differ so markedly from those of other species of the genus that its position in *Cerdistus* seems incorrect.

The spermathecae of *erythrurus* and related species form long, wide, thin-walled tubes of distinct form, tapering to a long, recurved end. The tubes narrow abruptly into narrow ducts which widen again before the valve. The valve is vase-shaped, cup-shaped, or conical, strongly sclerotized, and has a specific form in every species examined. Ejection apparatus very long, narrow, sclerotized, with external tubercles. Its muscles form a thick layer from above the valve to the common duct. The valve of *syriacus* is vase-shaped, and the reservoir is a large, ovoid sac which is darkly pigmented in fresh material. It is similar in a new species related to *syriacus*, but the valve is sclerotized only in its proximal, cup-shaped part and the reservoir has a short, finger-shaped process in its distal part. The processes on the ejection apparatus are longer near the valve in both species. Furca long, slender, divided as in *Neomochtherus*. Apodeme long and slender.

The spermathecae of *geniculatus* are long, thin, delicate tubes which do not form a coil. Valve truncate-conical, with transverse ridges. Furca divided, slender. End of lateral arms widened, curved.

Aedeagus of *C. erythrurus* of the *Neomochtherus* type, with a wide sheath and rudimentary prongs, denticles on the ventral side in the distal part, and two leaf-like, triangular apical processes which are relatively short in *erythrurus* and a related species but very long in another species in which the prongs are also very long. This species is probably identical with *Neomochtherus albicans oblitus* Tsacas, 1968, to judge by the drawing of the genitalia. Both *N. albicans* and *N. albicans oblitus* have an aedeagus closely resembling that of *C. erythrurus*, with leaf-like apical processes and a large dorsal basal plate, and clearly belong to *Cerdistus*. The basal part of the sheath of *erythrurus* and related species forms a large, rounded plate above the aedeagus. The aedeagus of *C. geniculatus*, *syriacus* and the related species is similar, but without the leaf-like processes or a basal plate.

Dististylus very long and slender, more or less curved, with specific differences.

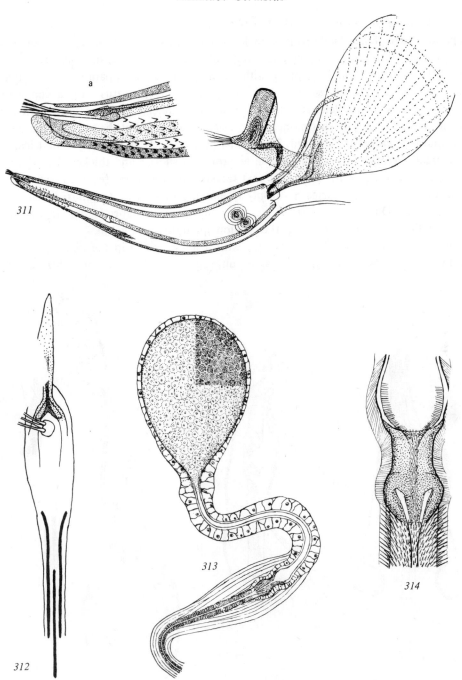

Figs. 311–314: *Cerdistus syriacus*
311. aedeagus; (a) end of aedeagus, enlarged; 312. furca of spermatheca;
313. reservoir of spermatheca, fresh material;
314. valve of spermatheca, fresh material

Eremisca autumnalis, heleni, osiris (Figs. 315–319)

The spermathecae of *autumnalis* form a thick, recurved, sclerotized tube. The sclerotization ends abruptly and the tubes narrow into a pointed funnel which is again sclerotized before the ring-shaped valve that is covered with tubercles. Ejection apparatus long, narrow, sclerotized, with fine tubercles. Proximal end of ejection apparatus with a cup-shaped sclerotization. Furca complete, elliptical, with a long apodeme and two sclerotized areas inside the lateral arms. The spermathecae of *heleni* are similar, but the tubes are narrower, longer, and less strongly sclerotized. The funnel before the valve is nearly cylindrical and the proximal sclerotization of the ejection apparatus is shorter and thicker. The reservoir of the spermathecae of *osiris* is shorter and thicker, its end is not recurved and its base is bulb-shaped, so that it forms a capsule resembling that of *Polyphonius*. Openings of canaliculi large. Ducts very short. Valve as in the other two species, but less strongly sclerotized. Proximal sclerotization of the ejection apparatus as in *heleni*.

Aedeagus of *autumnalis* has a wide sheath with a dorsal bulge in the distal part and a ventral bulge in the proximal part. Aedeagus pump very short. Prongs short, wide, function-

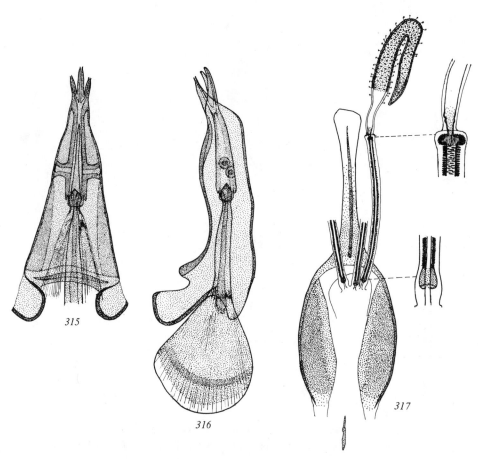

Figs. 315–317: *Eremisca autumnalis.* 315. aedeagus, dorsal;
316. same, lateral; 317. spermatheca

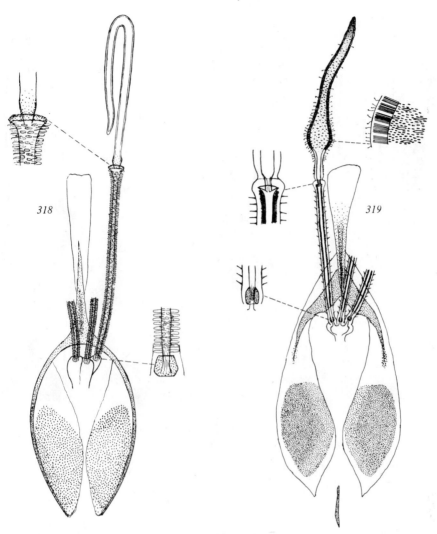

Figs. 318–319: Spermatheca. 318. *Eremisca heleni*; 319. *E. osiris*

al. Aedeagus of *heleni* with a wide, conical sheath. Pump very short, prongs short, wide, tapering, functional. Aedeagus of *osiris* similar, but the sheath has a large ventral bulge in the distal part, and the prongs are shorter. The aedeagus of all three species is connected with the sheath by two lateral columns in the basal half, as in *Neomochtherus*.

Gonopods rounded-triangular. Dististylus short, wide, with specific differences.

Echthistus rufinervis (Figs. 320–322)

The spermathecae form long, delicate tubes which are wide in the distal part and taper to a long, recurved end. Canaliculi distinct, with globular end, very numerous in the wide part of the tubes. Valve strongly sclerotized, broadly cup-shaped. Ejection apparatus short, sclerotized, with processes, ending in a part with longer processes and an internal

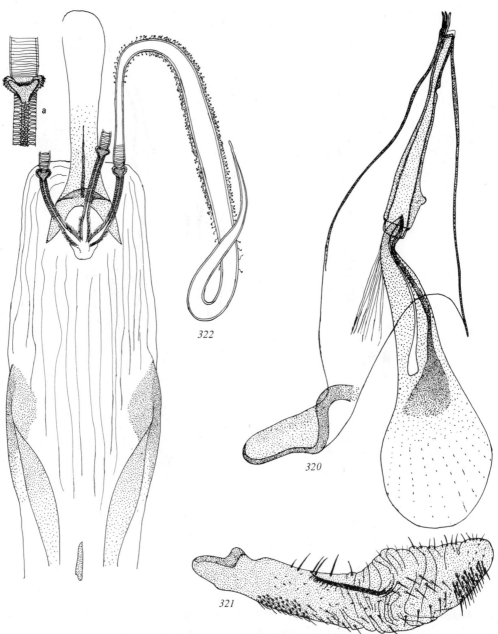

Figs. 320–322: *Echthistus rufinervis*
320. aedeagus; 321. dististylus; 322. spermatheca; (a) valve, enlarged

sclerotized tube. Common duct short and wide. Furca divided, apodeme wide, lateral
arms wide, weakly sclerotized.

Aedeagus of *Neomochtherus* type, sheath very wide, conical in lateral view, laterally
compressed, with a dorsal keel. Prongs very short, thin, non-functional. Apodeme wide.
Gonopods triangular, with three rounded, plate-shaped processes at the apex on the inner
side. Dististylus narrow at the base, wider and with transverse ridges apically.

134

Stilpnogaster aemulus (Figs. 323–325)

The reservoir of the spermathecae forms very wide, thin-walled sacs, densely covered with canaliculi with a large globular reservoir. The median spermatheca is larger and wider, tapering to a short, pointed process. The lateral spermathecae are narrower, with rounded end. Ducts short, covered with similar canaliculi, narrowing and then widening again before the valve. Valve broadly ring-shaped, ejection apparatus sclerotized near the valve, without processes. Furca divided, apical arc broad, apodeme narrow. Posterior median sclerite rod-shaped.

Aedeagus conical, sheath wide, with a distinct dorsal angle in the apical part. Prongs short, non-functional. Sheath without denticles. Apodeme long, moderately wide.

Gonopods oblong-triangular, with a slender apical process. Dististylus slender, with a rectangular process near the apex on the dorsal side.

Stilpnogaster was considered by Hull (1962) to be related to *Rhadiurgus*, which belongs to the *Machimus* group. However, the structure of the aedeagus with a wide sheath and rudimentary prongs, and the form of the spermathecae and furca suggest a closer relationship to the *Neomochtherus-Cerdistus* group.

Erax (*=Protophanes*) sp. (Figs. 326–327)

The spermathecae form long, wide, thin-walled tubes which taper to a long, thin, curled end. Ducts short, widening before the valve, which is cup-shaped, sclerotized and transversely striated. Ejection apparatus short, with fine processes. Furca slender, divided. Apodeme short, narrow.

Aedeagus of the *Neomochtherus* type, sheath wide, of nearly uniform width, slightly curved, tapering to a narrow end in the apical quarter. Prongs very short, non-functional. Sheath with narrow apical processes which extend beyond the prongs.

Gonopods long, curved, with rounded end. Dististylus slender, curved, with a club-shaped end with short spines.

New genus A (Figs. 328–329)

The species resembles *Neolophonotus* in the form of the face beard, the isolated prosternum, the bare postnotal callus and in the aedeagus without prongs. It differs from it in the reduced chaetotaxy of the thorax, so that it resembles *Neomochtherus* externally, and in the form of the ovipositor, which is laterally compressed and resembles that of species of *Machimus* or *Cerdistus*.

The spermathecae form spherical capsules with an indentation at the apex. Ducts short and narrow. Valve strongly sclerotized, nearly cylindrical. Ejection apparatus much longer than the ducts, narrow, with sclerotized tubercles which extend to part of the common duct. Furca very slender, nearly complete, lateral arms reaching to near the apical part.

Aedeagus conical, sheath wide, without prongs or other differentiations. The pump is connected with the sheath by two oblique, dorsal ridges in the basal part.

Gonopods short, triangular, with rounded end. Dististylus slender, parallel-sided, curved, with oblique, truncate end.

New genus B, four undescribed species (Figs. 330–339)

This genus also resembles *Neomochtherus* in external characters, but differs so markedly in the genitalia in both sexes that it must be considered as a new genus. Three of the species

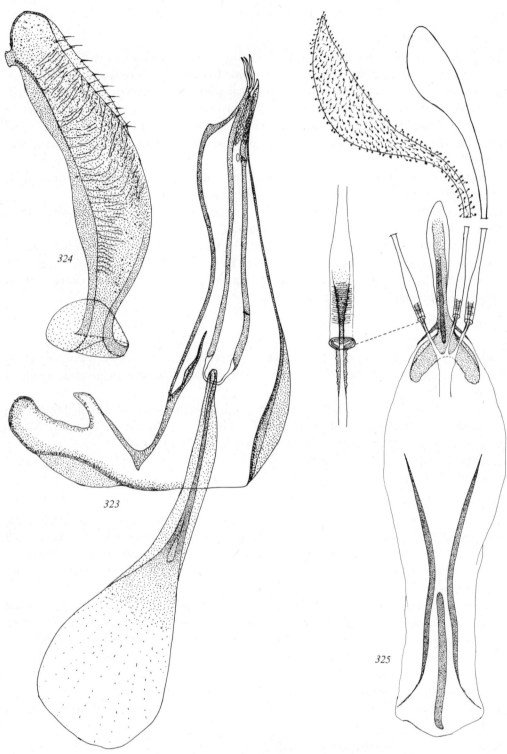

Figs. 323–325: *Stilpnogaster aemulus*
323. aedeagus; 324. dististylus; 325. spermatheca

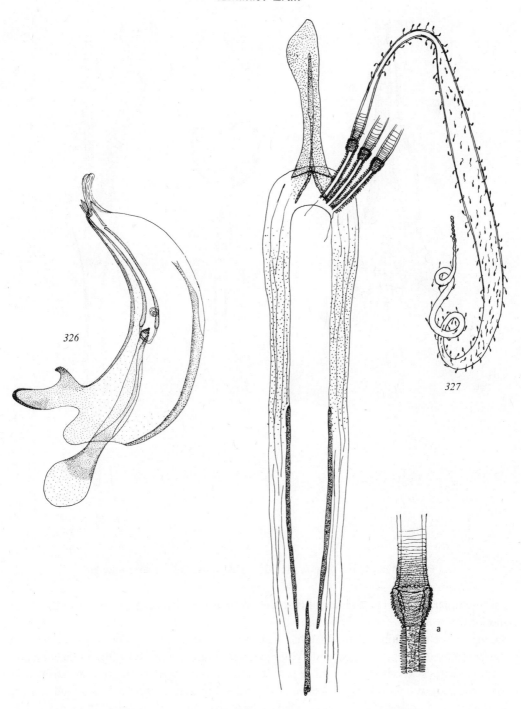

Figs. 326–327: *Erax* (= *Protophanes*) sp. 326. aedeagus;
327. spermatheca; (a) valve, enlarged

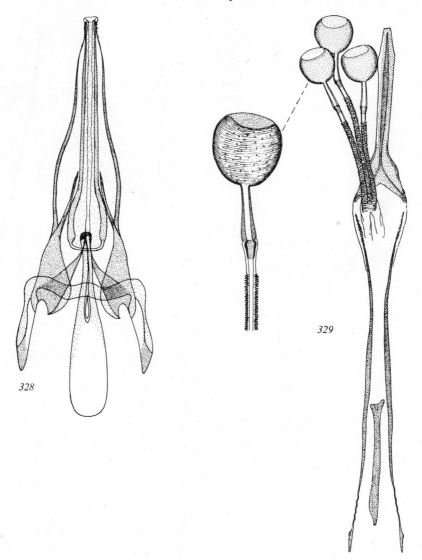

Figs. 328–329: Genus A. 328. aedeagus; 329. spermatheca

are so similar in external characters that they were at first thought to belong to the same species.

Spermathecae with thin-walled, relatively wide, short tubes of distinct form which taper to a short, recurved, slightly curled end. Valve conical, more or less sclerotized. Ejection apparatus short, sclerotized, with processes. It is wide and short in one species, striated in the distal part and with long processes at the base. Ejection apparatus longer and thinner in the other two species. Furca divided, apical part reduced, so that it consists only of the apodeme with a slightly wider base. Species no. 1 shows a differentiation of the common duct which has not been found in any other species. The common duct is very large and forms a distal bulge on which the ducts open. The basal part of the ventral wall

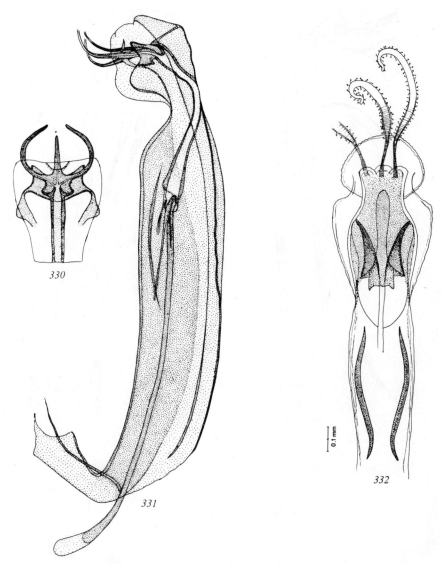

Figs. 330–332: Genus B, species no. 1
330. aedeagus; 331. end of aedeagus, dorsal; 332. spermatheca

and part of the lateral walls are strongly sclerotized, with a distinct marginal ridge. This sclerotization is absent in the other two species.

The aedeagus also shows complicated differentiations in the apical part. The sheath of species no. 1 is long and tubular; it narrows distally, and then widens again into an apical part with a sclerotized inner frame and three curved prongs. The lateral prongs are curved inwards. Aedeagus pump short, narrow, curved, apodeme long and slender. Aedeagus of species no. 2 similar, but with a large, rounded bulge before the narrow apical part on the dorsal side. Apical part not so large, trilobed, with short prongs. Aedeagus of species

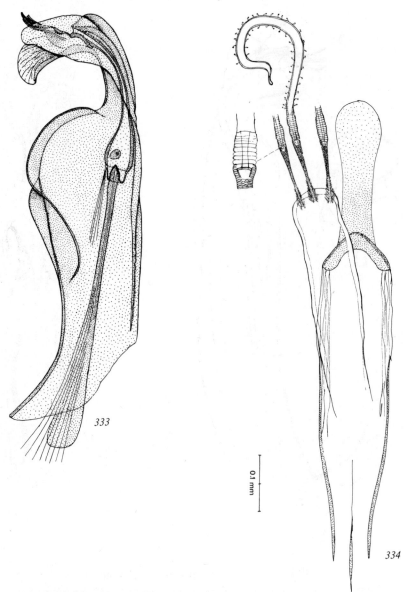

Figs. 333–334:   Genus B. 333. species no. 2, aedeagus;
334. species no. 3, spermatheca

no. 3 similar, but without a dorsal bulge, and the apical part smaller and of less complicated form than in the other two species.

Gonopods long, slender, triangular, with rounded end and with transverse ridges in two
of the species. Dististylus of characteristic form in the three species, with spines and transverse ridges (Figs. 337–339). Proctiger also of characteristic form. Sternite 8 of female of
characteristic form (Figs. 335–336).

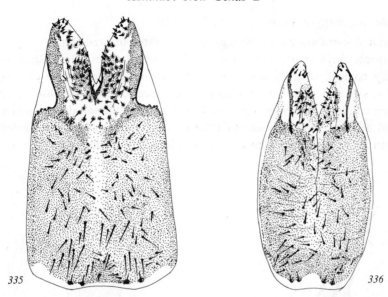

Figs. 335–336: Sternite of female, Genus B
335. species no. 1; 336. species no. 3

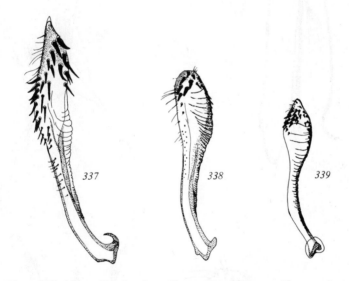

Figs. 337–339: Dististylus, Genus B. 337. species no. 1;
338. species no. 2; 339. species no. 3

Species no. 4 is provisionally placed in this genus. It resembles the other three species externally and in the form of the spermathecae, but differs markedly in the form of the aedeagus and dististylus. The aedeagus is triangular in lateral view, laterally compressed and has a large ventral keel and rudimentary prongs. The dististylus is short, crescent-shaped, with a lateral keel and without the characteristic spines of the other three species.

Philonicus albiceps, dorsiger, sinaiticus (Figs. 340–345)

The aedeagus and spermathecae of *albiceps* resemble those of *Eremisca*.

The spermathecae form short, wide, sclerotized, recurved tubes. Ducts short and wide. Valve broadly conical, strongly sclerotized. Ejection apparatus moderately long, thick, with fine processes and more strongly sclerotized processes near the base which form a network. They continue in membranous ducts and a long common duct. Furca complete, broad, lateral arms curved, with an inner plate-shaped extension. Apodeme long, wide, with a median ridge. Aedeagus conical with a wide sheath. Prongs short, wide, tapering,

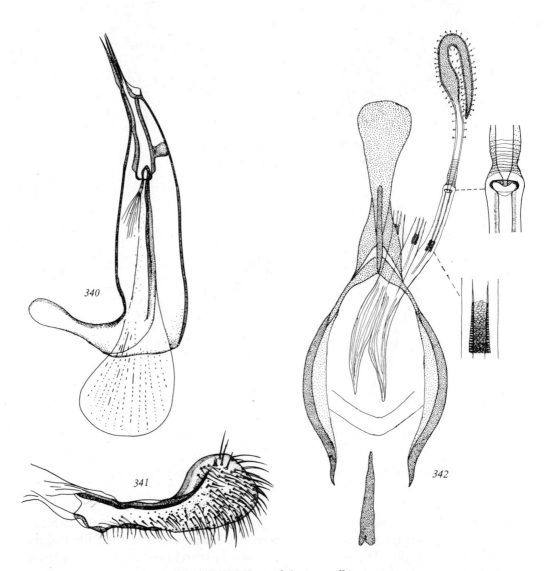

Figs. 340–342: *Philonicus albiceps*
340. aedeagus; 341. dististylus; 342. spermatheca

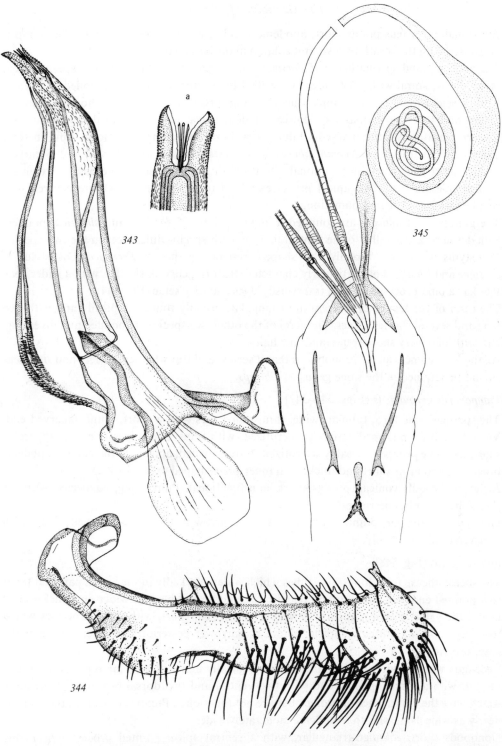

Figs. 343–345: *Philonicus sinaiticus* subsp.
343. aedeagus; (a) end of aedeagus, dorsal; 344. dististylus; 345. spermatheca

functional. Aedeagus pump short, apodeme slender, longer than the pump. The pump is connected with the sheath by a ventral column in the basal half.

The aedeagus and spermathecae of *dorsiger* and *sinaiticus* are different. The spermathecae form a loose spiral with very wide, thin-walled tubes tapering to a long, coiled end. Valve weakly sclerotized, ejection apparatus with fine processes, wide and short in *dorsiger*, longer and tapering in *sinaiticus*. Furca divided, apical part arc-shaped, apodeme long. Posterior median sclerite markedly bifid. Lateral arms widening posteriorly, bifid in some specimens. Aedeagus of *Neomochtherus* type. Sheath wide, conical, slightly S-curved. Prongs short and thin, non-functional. Sheath with fine denticles in the basal part on the ventral side and with two apical processes lateral to the prongs which are covered with small denticles. Aedeagus pump long.

The gonopods of *albiceps* are rounded-triangular, those of *dorsiger* and *sinaiticus* pointed, with dense ridges parallel to the ventral margin, with specific differences in the two species. Dististylus of *albiceps* broadly club-shaped, resembling that of *Eremisca*. Dististylus of *dorsiger* and *sinaiticus* of particularly characteristic form, curved, slender, with a wider head that has a bifid process and with transverse ridges and long setae (Fig. 344).

The cerci of the female of *albiceps* are triangular, broadly rounded apically and with two long and several shorter spines. The cerci of the other two species are pointed and diverging, and with only very short spines and fine hairs.

All these differences are so pronounced that it seems doubtful whether *dorsiger* and *sinaiticus* should be retained in the same genus as *albiceps*.

Pamponerus germanicus (Figs. 346–347)

The spermathecae form tubular, wide, thin-walled sacs with a short, blunt, recurved end. Valve broadly cup-shaped, strongly sclerotized, with transverse ridges. Ejection apparatus short, with fine processes, weakly sclerotized. Furca divided, apical sclerite curved, apodeme broad. Lateral arms short and thick, with inner plate-shaped extensions.

Aedeagus broadly conical, prongs short, non-functional. Pump long, connected with the sheath by two oblique ridges.

Gonopods triangular, pointed. Dististylus very narrow at the base, S-curved, wide, with transverse ridges apically.

Satanas gigas (Fig. 348)

The spermathecae form long, wide tubes which widen markedly in the apical part and taper to a pointed end. The ducts are crinkled in the middle and become wider and thick-walled before the valve. Ejection apparatus weakly sclerotized, short, with small processes which form regular rows. Furca complete, elliptical, narrower posteriorly. Apodeme wide at the base, tapering.

Aedeagus with a wide sheath, short, convex ventrally and with two large processes above its end, with a single opening and with two ventral and two dorsal processes at the apex which have the appearance of prongs, but are not functional. Pump very wide in the middle, nearly as wide as the sheath. Apodeme moderately wide.

Gonopods short, rounded-triangular, with a ventral apical pointed process. Dististylus slender, with numerous spines.

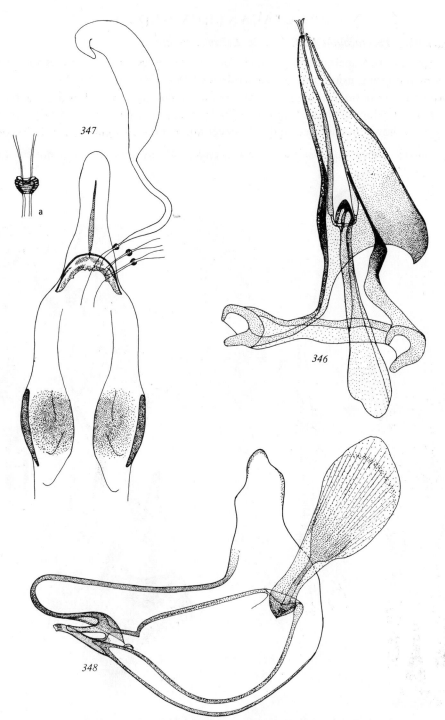

Figs. 346–347: *Pamponerus germanicus*. 346. aedeagus;
347. spermatheca; (a) valve, enlarged
Fig. 348: *Satanas gigas*, aedeagus

145

Proctacanthus, Proctacanthella, Eccritosia, Lecania, Myaptex

There are only two spermathecae, which form more or less ovoid capsules in the three genera of the group examined. Aedeagus divided into two long tubes to near the base. *Lecania* and *Proctacanthella* have not been examined, but are included in the group on the basis of the drawings of the aedeagus by Martin (1968). The numbers of the figures of the drawings are, however, confused, and it is not certain that the aedeagus is that of *Lecania*.

Proctacanthus brevipennis, longus, milberti, philadelphicus and an unidentified species (Figs. 349–354)

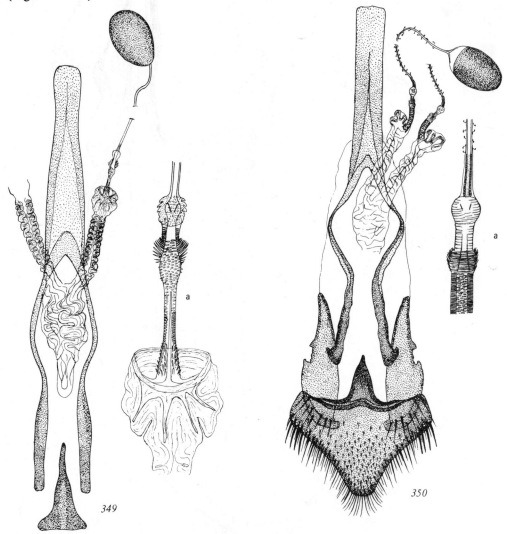

Fig. 349: *Proctacanthus philadelphicus*, spermatheca;
(a) valve and ejection apparatus, enlarged
Fig. 350: *Proctacanthus milberti*, spermatheca; (a) valve, enlarged

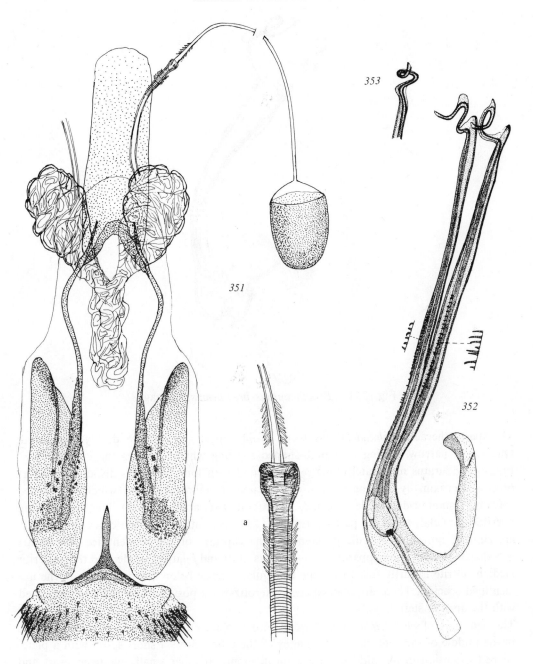

Fig. 351: *Proctacanthus* sp., spermatheca; (a) valve, enlarged
Fig. 352: *Proctacanthus milberti*, aedeagus
Fig. 353: *Proctacanthus longus*, end of aedeagus

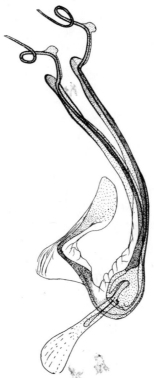

Fig. 354: *Proctacanthus brevipennis*, aedeagus

The spermathecae of *philadelphicus* form ovoid capsules which are darkly pigmented. Ducts very narrow, ending in a bulb-shaped swelling which contains the conical valve. Ejection apparatus moderately long, widened and with long processes distally and short processes proximally, ending in a wide, plicated, membranous duct which is widened apically. Common duct wide, plicated. Furca complete, apodeme long. Posterior median sclerite triangular, the wide part posteriorly. The spermathecae of the other four species are similar, but show specific differences. The capsules of the unidentified species are cup-shaped. The ejection apparatus of *milberti* is short and relatively wide, and the posterior median sclerite is partly fused with the triangular sclerite below the cerci (also in the unidentified species). There are plate-shaped sclerotizations posteriorly which are connected with the apex of sternite 8.

The aedeagus of *philadelphicus* is divided into two long tubes, with a row of denticles in the middle of the ventral side. The apex of the tubes forms a short spiral with a plate-shaped extension in its middle. Base conical, pump chamber small, apodeme short and narrow. Aedeagus of *longus* and *milberti* similar, but the spiral does not have a plate-shaped extension in *longus*. The aedeagus of *brevipennis* is markedly shorter, the spirals are longer, and ventral denticles are absent.

Eccritosia rubriventris (Figs. 355–356)

Capsules of spermathecae ovoid, darkly pigmented. Ducts long and narrow, covered with

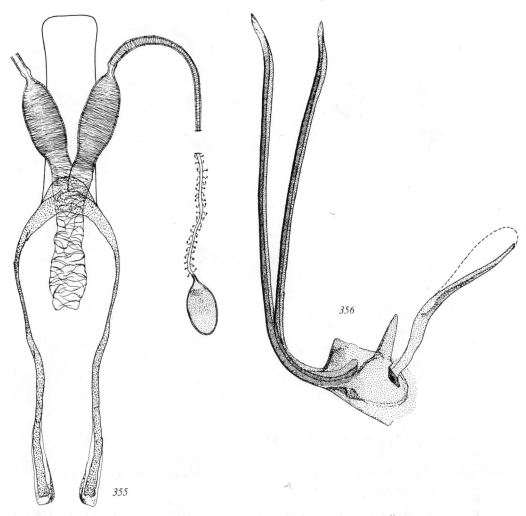

Figs. 355–356: *Eccritosia rubriventris*. 355. spermatheca; 356. aedeagus

canaliculi. Valve conical, weakly sclerotized, opening into a wide, striated sac and then into a plicated common duct. Furca complete, apodeme wide.

Aedeagus divided to near the base into two long tubes, each with a narrow sheath, and without a spiral at the apex. Pump chamber wide, rounded, larger than in *Proctacanthus*. Apodeme long, slender.

Myaptex brachyptera (Figs. 357–358)

Capsules of spermathecae ovoid, pigmented. Ducts very narrow, widening and membranous posteriorly. The valve consists of a cylindrical sclerotization with internal denticles. Furca narrow, complete, apodeme broad, short. A finger-shaped process at the inner curvature, which is also present in *M. hermanni* but absent in two other species illustrated by Artigas (1971). Posterior end of lateral arms of furca markedly widened, triangular.

Aedeagus divided to near the base into two long, thin, tapering tubes without a spiral or

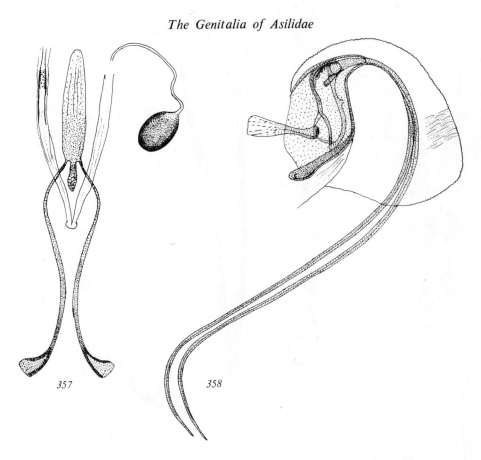

Figs. 357–358: *Myaptex brachyptera.* 357. spermatheca; 358. aedeagus

other differentiation. Base very large, pump chamber oblong, with spines in the ventral part. Apodeme very small, slender, triangular.

Efferia (=*Nerax* Hull) *aestuans, albibarbis, benedicti, pogonias* and an unidentified species (Figs. 359–376)

The spermathecae form capsules of varying form. They are narrowly oblong in *pogonias*, ovoid or nearly spherical in the other species examined. Ducts narrow, valve and ejection apparatus with distinct specific differences. Furca very long and narrow, extremely long in the species with a narrow, long ovipositor. Apodeme either short and broad or narrow. The furca may be complete, or the lateral arms are either narrowly connected with, or separated from, the apical arc, so that there are all transitions between a complete furca and a divided furca in the same genus. There are complicated differentiations in the apical part of the furca in *albibarbis*.

The aedeagus ends in three short, ventrally curved tubes, and the sheath of some species has two ventral processes near the apex which differ distinctly in form in the various species. They are long and curved in *pogonias*, triangular in *aestuans*, straight and with rounded end in *albibarbis*, and absent in the other species examined. The length of the aedeagus, of the aedeagus pump and of the apodeme also varies markedly.

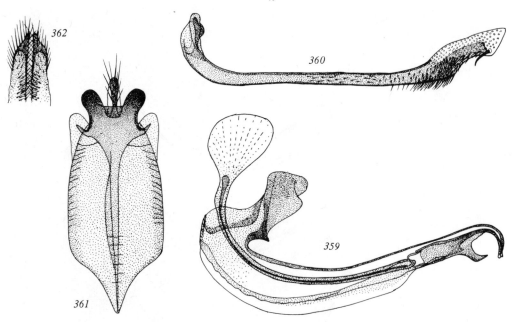

Figs. 359–362: *Efferia* (= *Nerax*) *pogonias*. 359. aedeagus;
360. dististylus; 361. proctiger, dorsal part; 362. same, ventral part

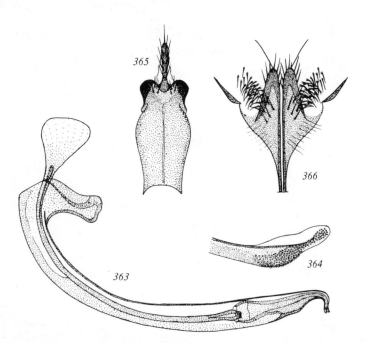

Figs. 363–366: *Efferia* sp. 363. aedeagus; 364. end of dististylus;
365. proctiger, ventral part; 366. same, dorsal part

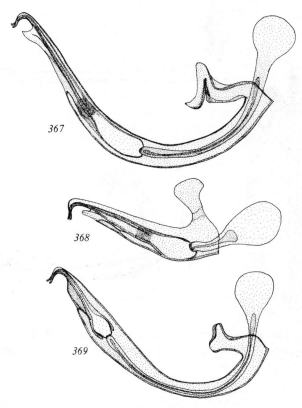

Figs. 367–369: Aedeagus of *Efferia*. 367. *E. aestuans*;
368. *E. albibarbis*; 369. *E. benedicti*

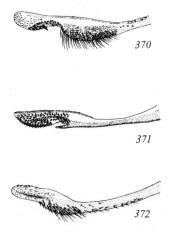

Figs. 370–372: End of dististylus. 370. *Efferia aestuans*;
371. *E. albibarbis*; 372. *E. benedicti*

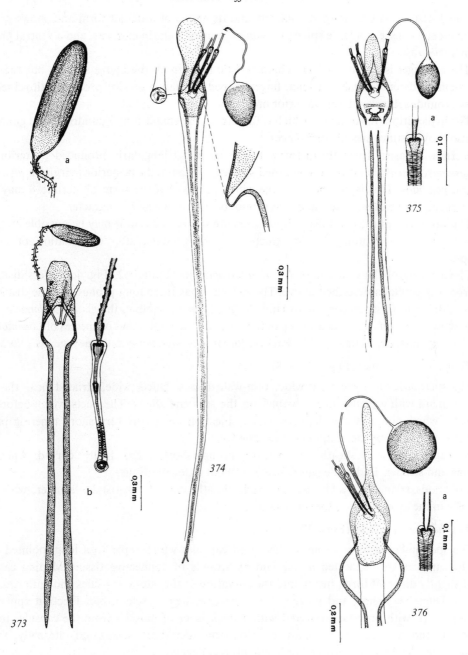

Figs. 373–376: Spermatheca of *Efferia*
373. *E. pogonias*; (a) capsule, enlarged; (b) valve
and ejection apparatus, enlarged; 374. *E. aestuans*;
375. *E. albibarbis*; (a) valve, enlarged; 376. *E. benedicti*; (a) valve, enlarged

153

The dististylus is very long and narrow and its end is of different form and shows specific differentiations, e.g. a large spine in *pogonias*, a small spine in *aestuans*, and a ventral process in *albibarbis*.

The proctiger is of similar, very characteristic form in all the above species, but there are specific differences. It has a long, finger-shaped median posterior process with setae, and two rounded, black, lateral posterior processes.

The hypandrium forms a ring which is fused with the epandrium. Epandrium and gonopods also show distinct specific differences.

Sternite 8 varies distinctly in form. It is long and triangularly produced posteriorly in *aestuans*, *benedicti* and the unidentified species, with straight posterior margin in *albibarbis* and *pogonias*. This is another example which proves that the form of sternite 8 may vary markedly in the same genus, and thus cannot be used as a generic character.

The variations of the genitalia in both sexes are so marked that it may be possible to establish subgenera according to the structure of the genitalia after examination of further species.

The drawing of the genitalia of *Nerax interruptus* in Hull (1962, Fig. 4c) differs distinctly from the genitalia described above. The aedeagus has three long prongs and the dististylus is of different form. According to Hull, *N. interruptus* resembles the genus *Lochmorhynchus* in some characters, and his drawing of the aedeagus resembles that of species of *Lochmorhynchus* illustrated by Artigas (1971). Parks (1968) places *interruptus* in the new genus *Triorla*.

Eccoptocus longitarsis (Figs. 377–378)

The spermathecae form very wide, thin-walled sacs. Ducts wide, striated near the sac. Canaliculi with globular head, distinct on the sacs and ducts. The ducts widen before the valve, which contains a wide, narrow ring. Ejection apparatus very short, tapering proximally. Furca complete, elliptical, apodeme wide.

Aedeagus with wide sheath. Pump short, prongs short, wide, slightly curved. Apodeme long and slender. Endoaedeagus delicate, without differentiations.

Gonopods triangular, with rounded end, slightly curved. Dististylus slender, with one ridge in the apical part on the inner side.

Polyphonius laevigatus (Figs. 379–381)

The spermathecae form strongly sclerotized capsules which taper to a long, pointed end. The capsules are connected at the end by strands of connective tissue. Median capsule of slightly different form, fitting into the curvature of the lateral capsules and with rounded end. Ducts very short and narrow. Valve narrow, weakly sclerotized. Ejection apparatus very short, with processes, covered with a thick layer of muscles from the capsules to the end. Gland restricted to the capsules. Common duct short, sclerotized internally. Furca complete, slender, posterior end of lateral arms widened.

Aedeagus with wide sheath, conical, straight. Pump very short, apodeme long and slender, longer than the pump. Prongs absent.

Gonopods of complicated form. Dististylus slender, with hook-shaped end and a bulge at the base.

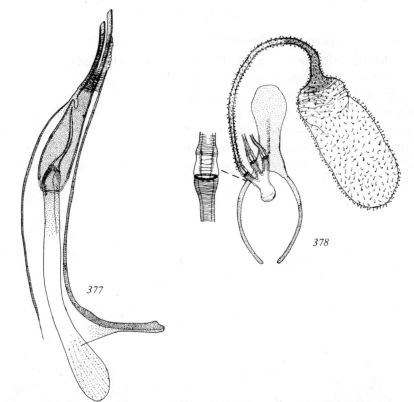

Figs. 377–378: *Eccoptocus longitarsis*. 377. aedeagus; 378. spermatheca

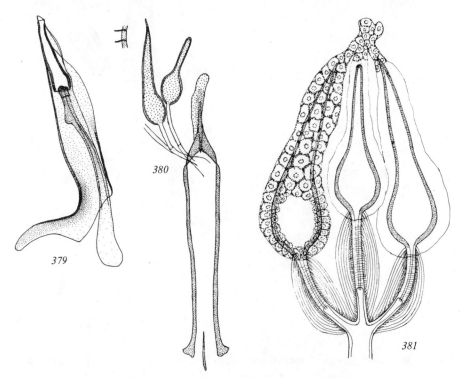

Figs. 379–381: *Polyphonius laevigatus*
379. aedeagus; 380. spermatheca; 381. same, fresh material

Neolophonotus porcellus and two unidentified species (Figs. 382–387)

porcellus. The spermathecae form wide, thin-walled tubes which end in wider, thin-walled sacs. Valve indistinct. Ejection apparatus striated, with processes at the base. Common duct long. Furca complete, short and wide, its apical part forming a triangle with the apodeme. Lateral arms with a posterior branch. Posterior median sclerite T-shaped.

Aedeagus with wide sheath, conical, with an angular process in the distal part on the ventral side. Pump short, apodeme long, slender. Prongs absent.

Gonopods with complicated processes and brushes of spines and setae. Dististylus short, with plate-shaped, rounded end.

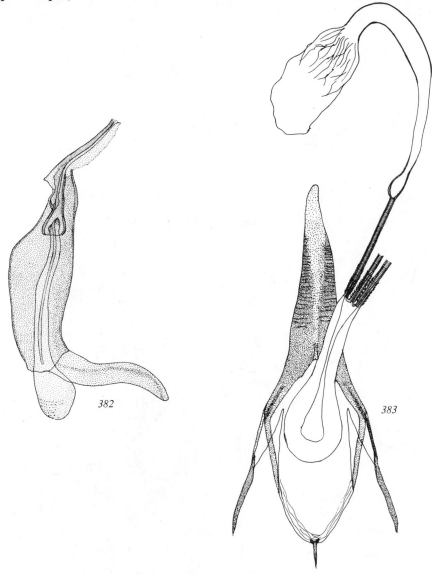

Figs. 382–383: *Neolophonotus porcellus.* 382. aedeagus; 383. spermatheca

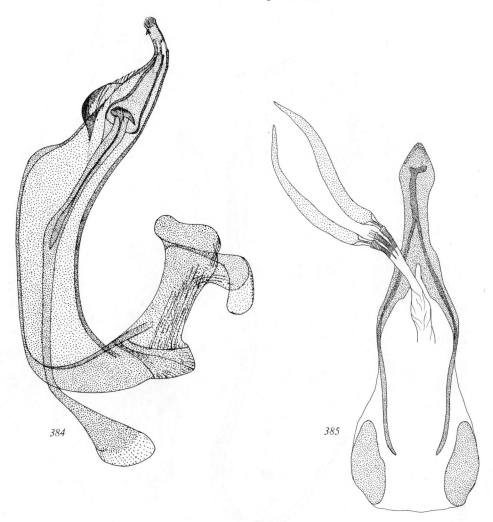

Figs. 384–385: *Neolophonotus* (*Lophopeltis*) sp. 384. aedeagus;
385. spermatheca

Species no. 1 (*Lophopeltis*). The spermathecae form thin-walled, wide, tapering tubes which narrow abruptly into a narrow, short, striated part. Furca complete, apodeme wide. Two plate-shaped sclerotizations lateral to the end of the lateral arms.

Aedeagus with wide sheath, pump very short, apodeme very long and curved. Sheath with a black bulge in the apical part on the ventral side. A short denticle near the end of the aedeagus. Prongs absent.

Gonopods and dististylus with complicated processes.

Species no. 2 (*Lophybus*). The reservoir of the spermathecae forms thin-walled, rounded sacs. Ducts long and thin, wider before the valve. Ejection apparatus striated. The three

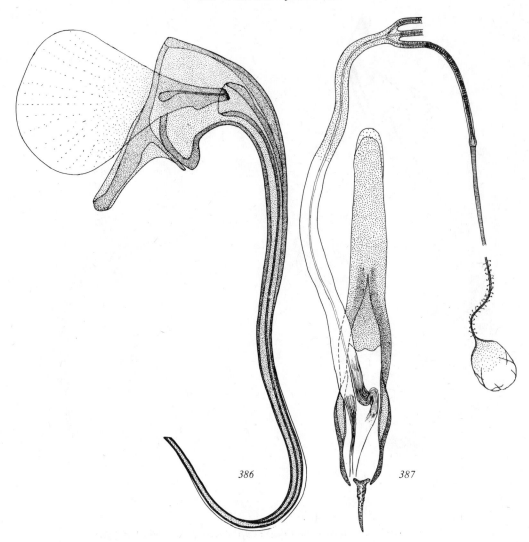

Figs. 386–387: *Neolophonotus* (*Lophybus*) sp. 386. aedeagus; 387. spermatheca

ducts unite in a very long common duct. Furca narrow, apodeme nearly as wide as the furca. Posterior median sclerite T-shaped.

Aedeagus very long and slender, simple, tapering, with recurved end. Apodeme broadly triangular.

Gonopods simple, short, truncate. Dististylus slender, parallel-sided, short, with a finger-shaped process and a broader, rounded process near the apex on the dorsal side.

The Old World genera of this group (*Promachus, Philodicus, Apoclea, Alcimus*) are characterized by the wing venation. There are usually 3 submarginal cells, R_4 extending proximally to R_{2+3}. However, this connection is incomplete in most species of *Apoclea* in which R_4 has only a recurrent stump or this is also absent, so that there are only 2 submarginal cells. Cell r_4 very long and narrow in the basal part, very wide apically. Prosternum isolated by membrane. Postnotal calli bare. Ovipositor tubular, not laterally compressed.

Promachus griseiventris, mustela, leoninus, the Ethiopian species *aequalis, fasciatus, poetinus, scotti, simpsoni* and an unidentified species (Figs. 388–397)

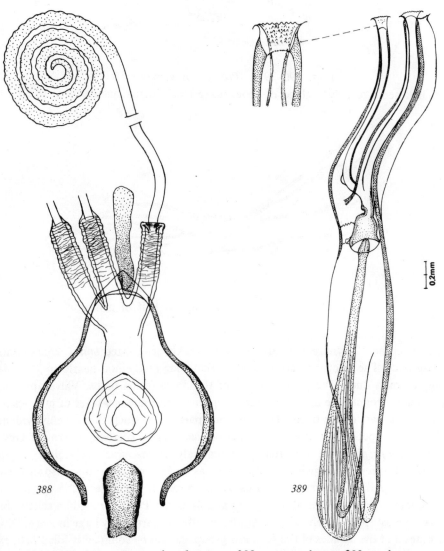

Figs. 388–389: *Promachus leoninus*. 388. spermatheca; 389. aedeagus

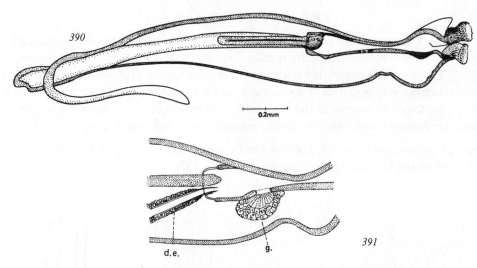

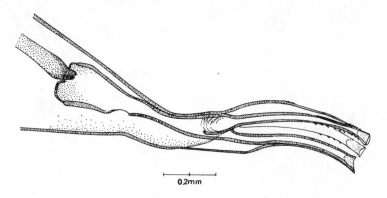

Figs. 390–391: *Promachus griseiventris*
390. aedeagus; 391. gland of aedeagus (g.) and ductus ejaculatorius (d.e.)

Fig. 392: *Promachus mustela*, apical half of aedeagus

The spermathecae of *griseiventris* form a loose, weakly sclerotized spiral which continues in a wide duct. The ducts are covered with glandular epithelium nearly to the opening into the common duct. The proximal end of the ducts forms bulbs. Valve and ejection apparatus not recognizable in the Palaearctic species, but there is a layer of muscles on the posterior, thick part of the ducts. The posterior part of the ducts is markedly widened in some Ethiopian species, forming wide, thick-walled sacs in *poetinus*, narrower sacs with processes in *simpsoni*. There is a broadly conical valve in *fasciatus* and *aequalis*, so that it seems justified to regard this wide posterior part as an ejection apparatus, also in species in which a sclerotized valve is absent. Furca violin-shaped in the Palaearctic species and in some Ethiopian species, elliptical in *aequalis* and *fasciatus*. Posterior median sclerite broad, triangular in most species, pointed posteriorly, of different form, very large in *scotti*. Oldroyd (1970) states that the females of the *fasciatus* group 'cannot be confidently identified, except by association with males'. The spermathecae of *aequalis* and *fasciatus* are similar, but there

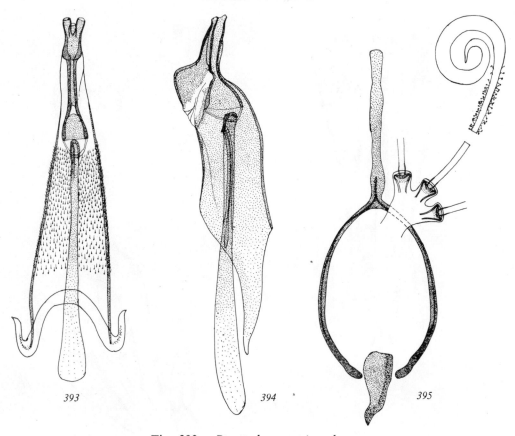

Fig. 393: *Promachus scotti*, aedeagus
Figs. 394–395: *Promachus fasciatus*. 394. aedeagus; 395. spermatheca

are specific differences, and the spermathecae may possibly provide characters for the distinction of the females. The spirals of the spermathecae of *leoninus* have a scalloped outline and the posterior part of the ducts is very wide, thick-walled and crinkled. Furca narrower anteriorly. Posterior median sclerite nearly rectangular. The spermathecae of *mustela* have spirals with fewer turns, and the posterior median sclerite is triangular and bifid posteriorly.

The aedeagus of the Palaearctic species examined ends in three tubes. The sheath is tubular, the pump short and the apodeme slender and longer than the pump. The tubes are short and wide in *griseiventris*, forming broad cups with internal denticles. The gland of the aedeagus opens in its distal part. The tubes of *mustela* are as long as the basal part, with a slightly funnel-shaped end and some denticles on the narrow neck. The aedeagus of *leoninus* is similar, but the tubes are wider, of uniform width, much longer and with a slightly differentiated end.

Gonopods short, fused in *griseiventris*. Dististylus short, wide, curved, with tapering end and numerous setae at the outer margin, with specific differences.

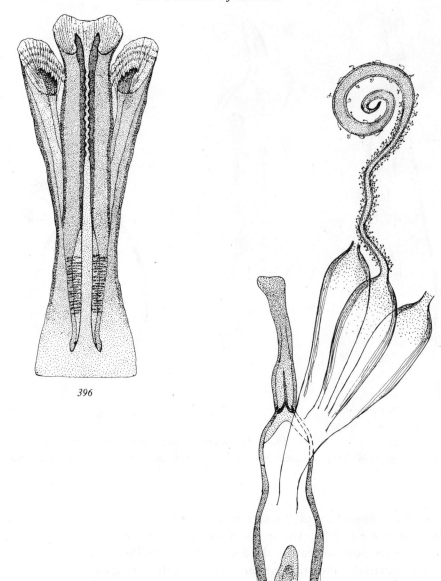

Figs. 396–397: *Promachus poetinus*. 396. end of aedeagus; 397. spermatheca

The form of the aedeagus varies widely in the Ethiopian species. There are species which have an aedeagus with a single, wide opening; species with three wide, cup-shaped tubes with markedly differentiated end and other variations. Numerous variations of the ending of the aedeagus and of the outline of the sheath have been illustrated by Hobby, unfortunately without sufficient details.

Apoclea algira, conicera, femoralis, helvipes, micracantha, trivialis and four undescribed species (Figs. 398–414)

The spermathecae form strongly sclerotized capsules. They are club-shaped, with a recurved, pointed process in most species examined, without such a process and darkly pigmented in *conicera*. They are nearly tube-shaped in *trivialis*. The ducts are shorter than the capsules in some species, but are longer in species nos. 3 and 4, in which the common duct is also much longer, the aedeagus ends in long tubes, and the apical part is not articulated by a membrane with the basal part. Canaliculi are recognizable only on the ducts, not on the capsules. Valve and ejection apparatus apparently absent. Common duct large, long, crinkled, membranous. Furca complete, lateral arms with widened posterior ends. Posterior median sclerite triangular in most species.

Aedeagus long and narrow, flattened in the basal half, with a funnel-shaped base. The apical part is sclerotized and articulated with the basal part by a membrane; it is folded back on the basal part at rest in most species. The three apical tubes are markedly differentiated in each species. They consist of an outer, wide sheath, through which passes an inner, narrow tube that opens in a sclerotized cup. The sheath is covered with denticles, the arrangement of which, as well as the form of the tube, differs in every species. The median tube differs from the lateral tube in some species. The aedeagus is of particularly complicated form in *trivialis*. The tubes resemble a heron's head, with a long beak and an S-curved neck. The membranous articulation of the apical part is absent in species nos. 3 and 4, in which the aedeagus ends in more or less long tubes with a differentiated end. These species resemble *Alcimus* in this respect.

There is a peculiar structure on the aedeagus of *Apoclea* and *Philodicus* which has not been found in the other genera of Asilini (Fig. 414). There are two ligaments which originate in cylindrical sclerotizations on the dorsal side of the aedeagus pump, near its base, and are inserted lateral to the head of the apodeme on the ventral posterior margin of the sheath. They are not muscles, as they do not stain with eosin or haematoxylin, and consist of fine fibres without nuclei or striation. They resemble the tonofibrillae at the insertion of some muscles. Their function is not clear.

Gonopods short, irregularly truncate, with specific differences. Dististylus short, of complicated form, with processes and setae; it is of distinctly different form in each species. Efflatoun (1937) considered *trivialis* a synonym of *micracantha* according to external characters, but a comparison of the genitalia showed that the two species differ distinctly and that *trivialis* is a valid species (Figs. 400, 405). *femoralis* was considered a synonym of *algira* by Engel (1930), but as a valid species by Efflatoun (1934). It differs distinctly from *algira* in external characters, in the male genitalia, and in the spermathecae.

Apoclea is thus a good example of the value of the genitalia in a genus in which external characters alone are not sufficient for the definition of species.

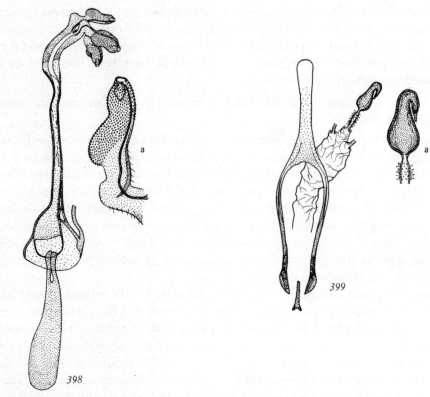

Figs. 398–399: *Apoclea algira*. 398. aedeagus; (a) terminal tube, enlarged;
399. spermatheca; (a) capsule, enlarged

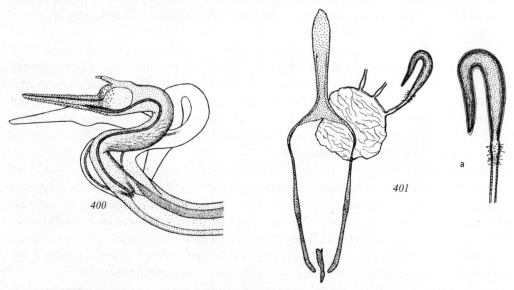

Figs. 400–401: *Apoclea trivialis*. 400. end of aedeagus; 401. spermatheca;
(a) capsule, enlarged

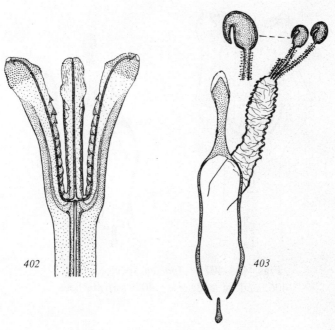

Figs. 402–403: *Apoclea helvipes*. 402. end of aedeagus; 403. spermatheca

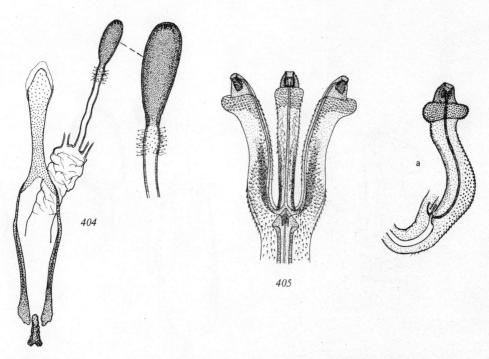

Fig. 404: *Apoclea conicera*, spermatheca
Fig. 405: *Apoclea micracantha*, end of aedeagus;
(a) one terminal tube, lateral

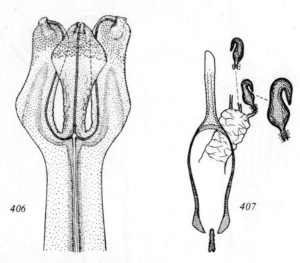

Figs. 406–407: *Apoclea* species no. 1
406. end of aedeagus; 407. spermatheca

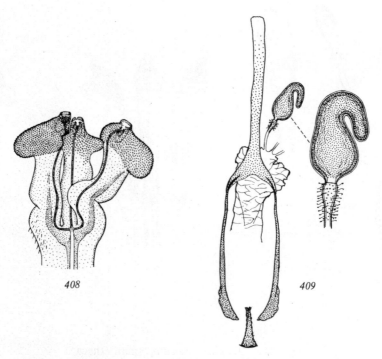

Figs. 408–409: *Apoclea* species no. 2
408. end of aedeagus; 409. spermatheca

166

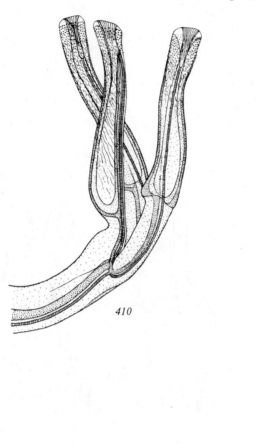

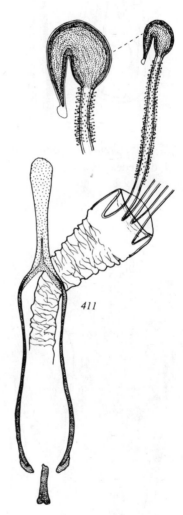

Figs. 410–411: *Apoclea* species no. 3
410. end of aedeagus; 411. spermatheca

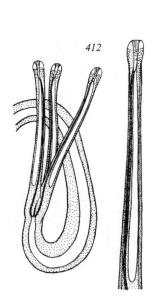

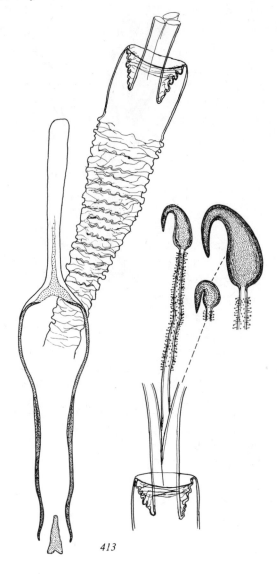

Figs. 412–413: *Apoclea* species no. 4
412. end of aedeagus;
413. spermatheca

413

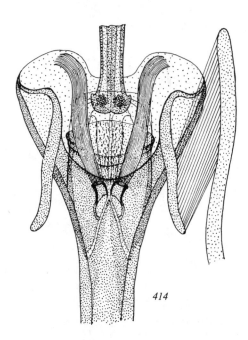

414

Fig. 414: *Apoclea* species no. 2,
ligaments at base of aedeagus pump

Philodicus ponticus, spectabilis and an unidentified species from South Africa (Figs. 415–424)

The spermathecae form club-shaped, straight or curved, strongly sclerotized capsules. Ducts narrow, long in *spectabilis*, shorter in *ponticus*, without differentiations. Furca complete, slender, apodeme wide.

Sternite 8 of the female of *ponticus* with two truncate apical processes and two slender, pointed lateral processes. That of *spectabilis* similar, but apical processes pointed, lateral processes wider.

ponticus. Aedeagus with a long, narrow basal part as in *Apoclea*. Apical part articulated by a membrane with the basal part and folded back on the basal part at rest, as in *Apoclea*. It ends in two wide tubes with striated margin. The median tube is finger-shaped and rudimentary, without an inner tube. The apical part bears a large dorsal process of complicated form. Apodeme short and narrow.

spectabilis. Aedeagus also with articulated apical part, which consists of three long, tapering tubes with a curled end which are covered with granules.

Gonopods and dististylus resembling those of *Apoclea* in general, but with distinct specific differences in the two species.

Philodicus sp. from South Africa. Aedeagus with broad, flattened basal part which widens only slightly at the base. Dorsal apodemes recurved posteriorly, rod-shaped. Apical part articulated by a membrane with the basal part and folded back on the basal part at rest. Apical tubes sclerotized, flask-shaped, with sclerotized inner tubes. Basal part with two cylindrical sclerotizations and with ligaments resembling those in *Apoclea*.

Alcimus sp. (Figs. 425–426)

The spermathecae form sclerotized tubes with a recurved, tapering end. Ducts long, their basal part wide, with scale-shaped differentiations with branched apical margin. Common duct very long, crinkled. Furca complete, slender.

Aedeagus very long and slender, with a wide, bulb-shaped base. It ends in three very long, thin tubes.

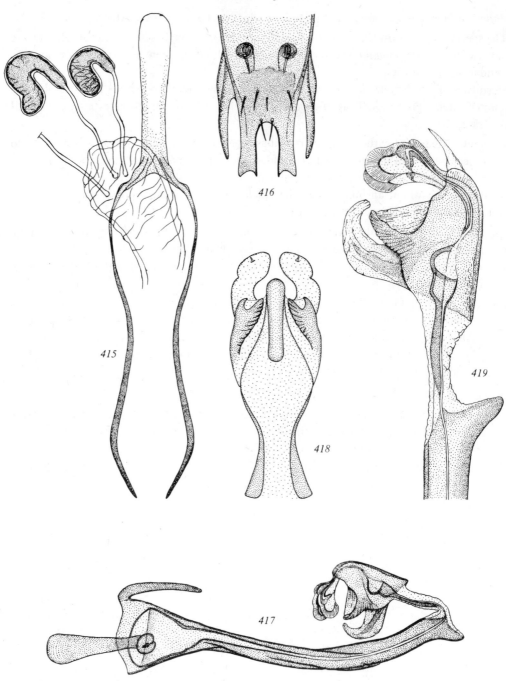

Figs. 415–419: *Philodicus ponticus.* 415. spermatheca;
416. end of sternite 8 of female; 417. aedeagus;
418. end of aedeagus, dorsal; 419. same, lateral

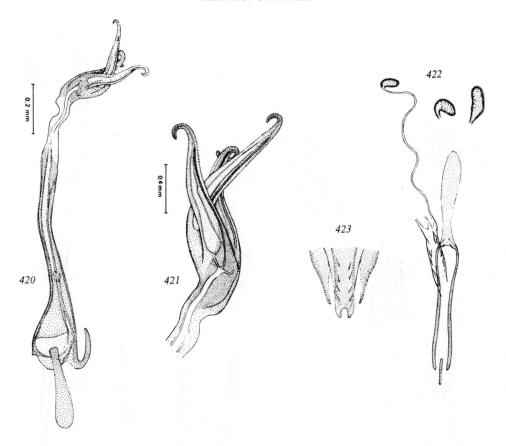

Figs. 420–423: *Philodicus spectabilis*
420. aedeagus; 421. end of aedeagus, enlarged;
422. spermatheca; 423. end of sternite 8 of female

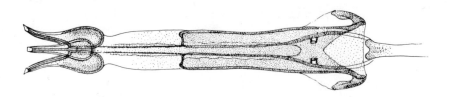

Fig. 424: *Philodicus* sp. (South Africa), aedeagus, dorsal

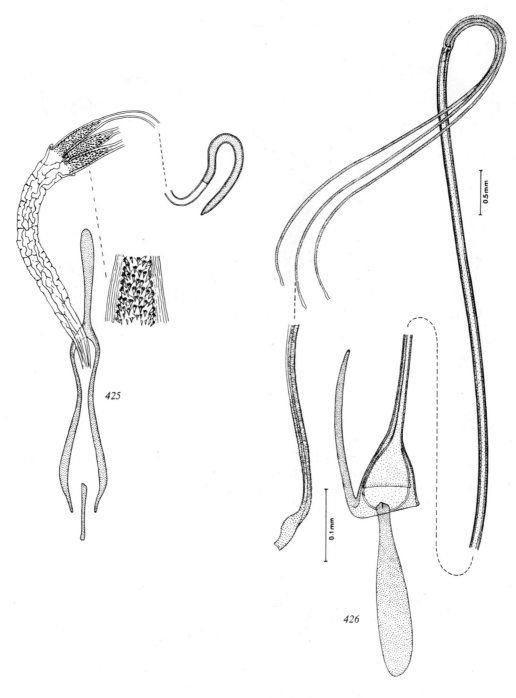

Figs. 425–426: *Alcimus* sp. 425. spermatheca; 426. aedeagus

CONCLUSIONS

THE STRUCTURE of the aedeagus and spermathecae described in this monograph shows that the genitalia contain important characters which make it possible in some cases to determine the systematic position of a species with more accuracy than external characters (*Cerdistus syriacus*, *Philonicus*, *Hoplistomerus*, *Trichardis* and others). The descriptions also show that a number of external characters which were considered as generic characters in the past are not valid (*Machimus* group).

The genitalia also contain characters which make identification possible in genera in which external characters are either so variable or so little different that identification according to external characters alone is practically impossible (*Apoclea*, *Habropogon* and *Stenopogon*). A more detailed description has therefore been given for some genera to illustrate the specific differences in the spermathecae and aedeagus.

The form of the aedeagus and spermathecae shows sub-family characters only in the Laphriinae. Aedeagus and spermathecae show generic characters in some genera (*Apoclea*, *Promachus*, *Saropogon* and others). The spermathecae alone show generic characters in *Habropogon*, in which the aedeagus is of the simplified type of the Dasypogoninae. It may become possible to define additional tribes in the Asilinae if the genitalia of further genera are examined. But the main importance of the genitalia will probably be in the identification of species.

Marked differences in the aedeagus and spermathecae were found in some genera which seem well defined externally (*Dysmachus*, *Eutolmus*). Only a few species of these genera were examined, however, and if other species are examined, it may be possible to divide these genera into subgenera or species groups.

The variation of the characters described could not be studied in sufficient detail from the material available, and some of the characters may prove to be within the range of variation of a species, particularly the relative length and the degree of sclerotization of the parts of the spermathecae. Still, the differences are so marked in most cases that they are apparently of specific rank.

It is premature to draw conclusions on classification from the material described; much more material from different zoogeographical regions will have to be examined first. But the great variety of the structures found in the limited material available suggests that the use of the detailed structure of the male and female genitalia will help to group species and genera more correctly in those groups in which external characters alone have proved inadequate.

This may also help to determine the distribution of species more exactly. Some species have been considered to have an extremely wide distribution. *Clinopogon sauteri*, described from Formosa, was recorded by Efflatoun (1937) from Egypt. This seems rather unlikely, and examination of the genitalia would almost certainly have shown that the Egyptian species is different from *C. sauteri* and perhaps more closely related to *C. maroccanus* Becker. Efflatoun also recorded *Saropogon pittoproctus*, described from Turkestan, from

Egypt. Comparison of *S. pittoproctus* from Turkestan with the similar species from Egypt and Israel showed that the two species differ distinctly in the genitalia.

While this monograph was in preparation, Artigas (1971) published a study of the aedeagus and spermathecae of the Asilidae of Chile. He began this study after I had drawn his attention to the wide variation of these structures. His conclusions closely agree with those in this work, particularly as to the generic and specific value of the differences involved.

BIBLIOGRAPHY

Abbassian-Lintzen R. (1964) 'Asilidae of Iran. II. Notes on the Genus *Eremisca* Zin. and Description of *E. shahgudiani*', *Ann. Mag. Nat. Hist.*, Ser. 13, Vol. 7; pp. 547–562.

Adler S. & O. Theodor (1926) 'On the *minutus* Group of the Genus *Phlebotomus* in Palestine', *Bull. Ent. Res.*, 16 : 399–405.

Artigas J. N. (1970) 'Los Asilidos de Chile', *Gayana (Zool.)*, No. 17, 472 pp.

— (1971) 'Las structuras quitinizadas de la spermatheca y funda del pene de los Asilidos y su valor sistematico a traves del estudio por taxonomica numerica', *Gayana (Zool.)*, No. 18, 106 pp.

Blasdale P. (1957) 'The Asilidae of the Genus *Philodicus* Loew in the Ethiopian Region', *Trans. Roy. Ent. Soc. Lond.*, 109 : 135–148.

Efflatoun H. C. (1934, 1937) 'A Monograph of Egyptian Asilidae'. I–II. *Mém. Soc. Roy. Ent. Egypte*, Vol. 4.

Engel E. O. (1930) in : Lindner, *Die Fliegen der Palaearktischen Region*. 24. Asilidae. Stuttgart, Schweizerbart.

Hobby B. M. (1936) 'The Ethiopian Species of the *fasciata* Group of the Genus *Bactria* (=*Promachus*), (Asilidae)', *Ent. Monthl. Mag.*, 72 : 182–199; 231–249; 274–299.

Hull F. M. (1962) 'Robberflies of the World', *U. S. N. M. Bull.* 224.

Jonescu M. A. & M. Weinberg (1961) 'Studien über die taxonomischen Merkmale und deren Variabilität bei einigen Raubfliegenarten (Asilidae)', *Rev. Biol.* (Romania), 6 : 425–434.

Karl E. (1959) 'Vergleichend-morphologische Untersuchungen des männlichen Kopulationsorgans bei Asiliden', *Beitr. Entom.*, 9 : 619–680.

Lyneborg L. (1968) 'Notes on Two Species of *Machimus* Loew in Northern Europe', *Notulae Entom.*, Helsingfors, 48 : 131–135.

Mackerras I. M. (1955) 'The Classification and Distribution of the Tabanidae. III. Morphology', *Austr. J. Zool.*, 3 : 444–447.

Martin C. H. (1968) 'The New Family Leptogastridae', *J. Kansas Entom. Soc.*, 41 : 70–100.

Oldroyd H. (1946) 'Notes on Some Asilidae Taken in Corsica with Description of a New Species', *Encycl. Entom. Diptera*, 10 : 27–31.

Oldroyd H. (1954) 'The Horseflies (Tabanidae) of the Ethiopian Region', II, *Brit. Mus. (Nat. Hist.).*

— (1963) 'The Tribes and Genera of the African Asilidae', *Stuttg. Beitr. Naturk.*, No. 107, 16 pp.

— (1970) 'Studies on African Asilidae. I. Asilidae of the Congo Basin', *Bull. Brit. Mus. (Nat. Hist.)*, 24 : 209–334.

Ovazza M. & R. Taufflieb (1954) 'Les Genitalia Femelles des Tabanides et leur Importance en Systématique', *Ann. Parasitol.*, 29 : 250–264.

Owsley W. B. (1946) 'The Comparative Morphology of Internal Structures of the Asilidae', *Ann. Ent. Soc. Amer.*, 39 : 33–68.

Parks L. (1968) 'Synopsis of Robberfly Genera Allied to *Efferia* and *Eicherax*, Including a New Genus', *Pan-Pacific Entom.*, 44 : 171–179.

Reichardt H. (1929) 'Untersuchungen über den Genitalapparat der Asiliden', *Zeitschr. wissensch. Zool.*, 135 : 257–301.

Snodgrass R. E. (1935) *Principles of Insect Morphology*, McGraw-Hill, New York.

Theodor O. (1965) 'On the Classification of American Phlebotominae', *J. Med. Entom.*, 2 : 171–197.

Tsacas L. (1964) 'Revision des espèces du genre *Acanthopleura*', *Mém. Mus. Nat. Hist. Natur. sér. A. (Zool.)*, 28 : 205–240.

— (1967) 'Un nouvel *Acanthopleura* de l'île de Rhodes', *Verh. Naturf. Ges. Basel*, 78 : 311–314.

— (1968) 'Revision des espèces du genre *Neomochtherus*. I. Région Paléarctique', *Mém. Mus. Nat. Hist. Natur.*, 47 : 127–328.

Verrall G. H. (1909) *British Flies*, 5 : 679.

Weinberg M. (1967) 'Further Data on the Asilidae from the Danube Basin', *Trav. Mus. Nat. Hist. Grigore Antipa*, 7 : 299–311.

Wesché W. (1906) 'The Genitalia of Both Sexes of Diptera and their Relation to the Armature of the Mouth', *Trans. Linn. Soc. Lond.*, Vol. 9.

נדפס בדפוס ליתר־אופסט זיו, ירושלים

כתבי האקדמיה הלאומית הישראלית למדעים

החטיבה למדעי־הטבע

———

על מבנה חלקי המין של אַסִילִידֶה

מאת

א׳ תאודור

ירושלים תשל״ו